Business and Legal Forms

for

Industrial Designers

Tad Crawford, Eva Doman Bruck and Carl W. Battle

NEW ENGLAND INSTITUTE
OF TECHNOLOGY
LIBRARY

iDSA

Allworth Press, New York   **Industrial Designers Society of America**

10-08

#57003807

© 2005 Tad Crawford
© 2005 Eva Doman Bruck
© 2005 Carl W. Battle

All rights reserved. Copyright under Berne Copyright Convention, Universal Copyright Convention, and Pan American Copyright Convention. No part of this book may be reproduced, stored in a retrieval system, or transmitted in any form, or by any means, electronic, mechanical, photocopying, recording or otherwise, without prior permission of the publisher.

Published by Allworth Press, an imprint of Allworth Communications, Inc., 10 East 23rd Street, New York, NY 10010.

Cover design by Derek Bacchus
Book design by SR Desktop Services, Ridge, NY
Typography by Integra Software Services, Pondicherry, India
Printed in Canada

Library of Congress Catalog-in-Publication Data

Crawford, Tad, 1946—
    Business and legal forms for industrial designers/Tad Crawford, Eva Doman Bruck, Carl W. Battle.
        p. cm.
    Includes index.
    ISBN: 1-58115-398-8
1. Design, Industrial—Law and legislation—United States—Forms.   2. Contracts for work and labor—United States—Forms.   I. Bruck, Eva Domain.   II. Battle, Carl W.   III. Title.

KF2930.I54C73 2004
343.73'0787452—dc22

2004025943

This book is designed to provide accurate and authoritative information with respect to the subject matter covered. It is sold with the understanding that the publisher is not engaged in rendering legal, accounting, or other professional services. If legal advice or other expert assistance is required, the services of a competent attorney or professional person should be sought. While every attempt is made to provide accurate information, the author and publisher cannot be held accountable for any errors or omissions.

Contents

About the Authors

Tad Crawford has served as general counsel for the Graphic Artists Guild and legislative counsel for artists' groups fighting for greater rights. An attorney and author of *Legal Guide for the Visual Artist* and many other books for creative professionals, he is also the legal affairs editor for *Communication Arts* magazine.

Eva Doman Bruck has been a design industry professional for twenty-five years and has managed the business and legal aspects of interdisciplinary design projects for Time Warner's Digital Production Studio, Milton Glaser, Inc., and other companies. She is a faculty member of the MFA in Design program at the School of Visual Arts and co-author of *Business and Legal Forms for Graphic Designers* (Allworth Press). A manager for the Hearst New Media Center, she lives in New York City.

Carl W. Battle is an attorney and author who has dealt with intellectual property in the realm of industrial design throughout his career. He has secured patents and trademarks for products including packaging, medical goods, consumer goods, and parts of a space shuttle, for leading corporations such as B.F. Goodridge, Sandoz, Schering Plow, and Pharmacia. He is the author of *The Patent Guide* and *Legal Forms for Everyone* (Allworth Press) and lives near San Francisco, California.

A System for Success

The knowledge and use of good business practices is an essential step toward success for any professional or company, including the industrial designer and the design firm. The forms contained in this book deal with the most important business transactions that a designer or design firm is likely to undertake. At the back of the book is a CD-ROM containing all the forms, which will allow the designer to customize and easily revise them. The fact that the forms are designed for use, and favor the designer if negotiations are necessary, give them a unique value.

The Process of Industrial Design

Industrial design is that part of product development that focuses on the ornamental or aesthetic character of a product. The design might focus on three-dimensional features, or the physical aspect of an object; it might also be directed to a two-dimensional, or surface design. Applications can range from high-tech industrial products to handcrafted artistic objects in just about any industry. An industrial designer might be an individual practitioner or a member of a larger, interdisciplinary organization comprised of engineers, technologists, machinists, marketers, and salespeople, among others.

In general, the profession requires creative vision; a substantial understanding of end-user behavior; continuous tracking of market conditions; effective company-wide communications; and considerable technical and manufacturing knowledge (or the ability to employ it). For the designer-entrepreneur, there is the additional demand to find clients and/or market products, handle legal procedures, and to manage on both micro (project) and macro (business) levels. In many instances, industrial designers drift into the realm of product development. This book is more focused on the traditional practice of industrial design, but acknowledges the product development cycle as appropriate.

While a range of design solutions or original creations may vary from the decorative to the highly technical, the *process* of their creation is remarkably similar.

The creative endeavor ideally begins with an exploration of the necessity for a product, the characteristics of its end-user audience, and the business environment in which it will live, including an assessment of its various competitors. Preplanning is the critical stage during which the development of the product is orchestrated from a strategic point of view. This is when the parameters of functional scope, physical form, cost of design and development, manufacturing, marketing/sales opportunities, distribution and maintenance services must be considered. Decisions are made at this stage concerning allocations of time, money, material and human resources, including the selection of project team members. Organizations that establish cross-company teams for product development and require early and frequent participation by members of the research, design, assembly, sales, and service units, tend to deliver products far more efficiently and reliably than those that isolate these functions. Useful and timely input helps identify and prevent time and cost-consuming errors throughout the development process (see new product decision matrix, Form 6).

The creative process continues into ideation, or concept exploration, resulting in sketches, computer-generated and/or handmade models in clay, wood, foam, or other easily manipulated materials. Design development includes the study of key alternative ideas and their refinement, concluding in the selection of the one or two designs that will be prototyped, tested, further refined, tested again, and eventually specified for production. It cannot be overemphasized that recognizing mistakes

INDUSTRIAL DESIGN PROCESS

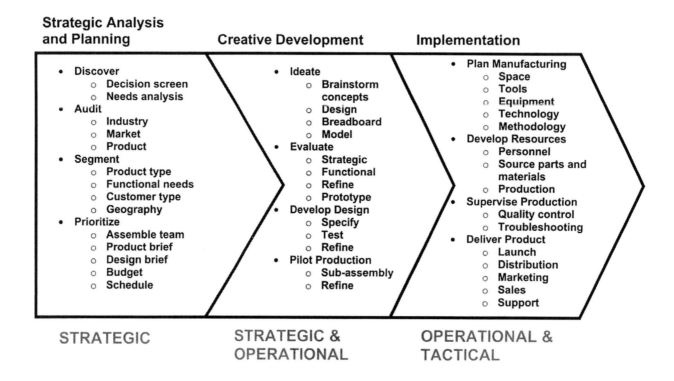

and correcting them early in the development process is least costly and most time effective.

While an industrial designer might be like other design professionals—hired by employers and clients—it is also common among industrial designers to act as entrepreneurs who bring their own products to life in the marketplace. Regardless of whether you are working for an employer, a client, or yourself, the necessity of bringing a product to fruition profitably and on time requires that there be standard ways of processing workflow and an organized system for keeping track of data and communicating information. Additionally, it is of critical importance to assess the product's legal and safety requirements as well as to protect the confidentiality of designs throughout the creative and production processes in order to secure legal ownership and usage rights.

Using the Forms

The purpose of *Business and Legal Forms for Industrial Designers* is, in part, to provide information, systems, and forms that are useful to the organization and smooth functioning of both the creative and business sides of an industrial design practice. The industrial design profession lends itself to a variety of scenarios—from individuals who work as employees for in-house design departments focusing strictly on the design and development of products, to those who are responsible for the entire scope of product development, to those who offer discrete areas of expertise to clients strictly on a consulting basis.

Industrial designers and product development consultants will find many of these forms useful, but this book is also intended to serve the needs of individuals who run businesses, departments, and/or large-scale projects. This book encompasses the *management* of individual projects as well as the business entity itself. The forms are designed to serve the purposes of the business owner, independent contractor, the employee of an industrial design firm, or an industrial design department of a larger organization. Again, the forms provided in this book are intended to be flexible—to be altered, used as a whole system, or in parts—as they best suit different circumstances. For the purposes of clarity, in instances where we refer to "clients," we understand that the client may very well be you, the designer, since many industrial designers are entrepreneurs and therefore their own clients.

The order of the forms is based upon the general chronology of events that occur in the course of running a project (with forms for the more sophisticated contracts coming last). In some areas, however, forms dealing with specific activities may be grouped together. To start, the *job index* is the point of origin because it provides every assignment with a job number. Some firms prefer to assign job numbers only to projects that have committed budgets. Because the preplanning needed for an industrial design project is extensive and the decision to commit to financing a project may not occur until there is some level of assurance of viability, a marketing number or an advance-of-approval number could be assigned. Specific hours and/or specific budgets are usually allotted to a pre-project number. At this stage, a very extensive assessment is needed to ascertain the feasibility of the proposed product—not only from a purely functional perspective, but also from the point of view of its fit with the company's profile, the industry, and the marketplace. For this purpose, we've included a *new product decision matrix*.

Following the opening of a job number, a *job sheet* is started where all project details are captured, including ongoing time and costs. As soon as either a product is selected for development, or a prospective client asks for an estimate or a bid and invites the designer to attend meetings and/or submits documents that reveal enough information about the proposed assignment, it is possible to formulate a *project plan and budget*. The difference between an estimate (an approximate calculation) and a bid (a specific proposal) is that the numbers in a bid are binding (as long as specifications and schedules are unchanged). In the estimating format, this particular form can be used as a worksheet to calculate estimated costs and time; in the summary format, it can also be submitted as part of the official estimate. With the addition of two columns, it can also serve as a status and variance report.

The *credit reference* form is used to ensure that an unknown client has a satisfactory history of bill payment. With the budget and schedule in place, the designer is ready to assemble a comprehensive proposal with the *proposal* form, as well as a *comprehensive schedule*. *Time sheets* are used to record staff time expended on both billable and non-billable activities.

A very useful form is the *status report* for either the client or other inter-company units. It's an easy way to communicate a product's progress at regular intervals, as well as to highlight issues that need to be addressed. An extremely important form is the *work change order*. Using it can mean the difference between losing money and making a profit on a project. It is the safeguard against capricious and indecisive clients—or management. A formal assurance of client agreement and an important step in the process is the *approval* form.

Anytime it is necessary to send documents, samples or any other goods, a *transmittal* form is used to identify the sender, the addressee and the items being transported, whether by hand, mail, courier, or facsimile. To keep a handle on the various items in transit at any one time, especially when using more than one form of shipping, use the *traffic log* form.

To keep track of specifications, bids and agreements with sub-contractors, the *contractor log* form is handy. It can also serve as the one-stop record of samples, photos and detailed drawings needed to keep a handle on any outsourced elements. To simplify the process of soliciting vendors' bids and/or estimates, making sure to use the same specs with competing vendors, use the *estimate request* form. The *estimate log* can then be used to keep track of competing estimates and/or bids coming from vendors. Once decisions are finalized as to which vendors to use, the *purchase order* is used either by itself, or with a formal contract, to place orders for selected goods and services. *Expense reports* are used to organize out-of-pocket purchases and/or travel receipts incurred by a project.

We've included a set of forms for keeping track of payables and receivables. It is not suggested here that these forms replace the necessity of abiding by generally accepted accounting practices (GAAP in accounting lingo) and the maintenance of standard bookkeeping ledgers. The forms in this book are meant to serve as a very simplified method of keeping track of financial information. The *payables index* provides a way of tracking all incoming bills as well as a system for having these bills reviewed and identified for approval, billable or not. The *billing index* provides invoice numbers as well as a handy guide for reviewing the status of payments due. Two types of invoices are included in this section, as well as a *statement* form and a *final notice*.

Even the most successful industrial design firms continue to market themselves, no matter how busy they are with current work. The *marketing log* and *qualifying checklist* are used to track the constant effort of making sure that there is a constant flow of prospective new projects.

The forms then shift to focus more on contractual relationships and intellectual property. A crucial form, the *industrial designer–client agreement*, gives a format in which to address the many different issues that are likely to arise. Since office space is likely to be a concern, there are forms for leasing, subletting, and assigning premises. A contract is given for use with the fabricator of a design as well as another contract for use with independent contractors. There is a *confidentiality agreement* for the submission of ideas or designs as well as a contract to license a design to others for manufacture. There are a number of

employment forms, including an *employment agreement* and a *restrictive covenant* with respect to employment. And intellectual property issues are covered in the areas of copyright, trademark, and utility and design patents. Because so many of these forms are contractual, it is worthwhile to focus on some salient features of contracts.

Handling Contracts

A contract is an agreement that creates legally enforceable obligations between two or more parties. In making a contract, each party gives something of value to the other party. This is called the "exchange of consideration." Consideration can take many forms, including the giving of money or a consultation or the promise to consult or pay for a consultation in the future.

All contracts, whether the designer's or someone else's, can be changed. Before using the contracts in this book, the designer should review them with his or her attorney. This gives the opportunity to learn whether local or state laws may make it worthwhile to modify any of the provisions. For example, would it be wise to include a provision for arbitration of disputes, or are the local courts speedy and inexpensive to use, so no arbitration provision is necessary?

The contracts must be filled out, which means that the blanks must be completed. Beyond this, however, the designer can always delete or add provisions on any contract. Deletions or additions to a contract are usually initialed in the margin by both parties. It is also a good practice to have each party initial each page of the contract, except the page on which the parties sign.

The designer must ascertain that the person signing the contract has authority to do so. If the designer is dealing with a company, the company's name should be included, as well as the name of the individual authorized to sign the contract and the title of that individual (or, if it isn't clear who will sign or that person has no title, the words "Authorized Signatory" can be used instead of a title).

If the designer won't meet with the other party to sign the contract, it would be wise to have that party sign the forms first. After the designer gets back the two copies of the form, they can be signed and one copy returned to the other party. As discussed in more detail later under Letter Contracts, this has the advantage of not leaving it up to the other party to decide whether to sign, and thus make a binding contract.

If additional provisions that won't fit on the contract forms should be added, simply include a provision stating, "This contract is subject to the provisions of the rider attached hereto and made a part hereof." The rider is simply another piece of paper, which would be headed, "Rider to the contract between _____and _____ dated the _____ day of _____, 20_____." The additional provisions are put on this sheet, and both parties sign it.

Negotiation

Understanding the business concepts behind the forms is as important as using them. By knowing why a certain provision has been included and what it accomplishes, the designer is able to negotiate when faced with someone else's contract. The designer knows what is and is not desirable.

Contracts require negotiation. The forms in this book are favorable to the designer. When they are presented to a client, fabricator, or contractor, changes may very well be requested. The explanation in this book of how to use each form should help the designer evaluate changes that either party may want to make. The explanation should also help the designer understand what changes would be desirable in forms presented to the designer.

Keep in mind that negotiation need not be adversarial. Certainly, the designer and the other party may disagree on some points, but the basic transaction is something that both want. This larger framework of agreement must be kept in mind at all times when negotiating. Of course, the designer must also know which points are nonnegotiable and be prepared to walk away from a deal if satisfaction cannot be had on these points.

When both parties have something valuable to offer each other, it should be possible for each side to come away from the negotiation with a winning feeling. This win-win negotiation requires each side to make certain that the basic needs of both parties are met, so that the result is fair. The designer can't negotiate for the other side, but a wise negotiation strategy must allow the other side to meet its vital needs within the larger context that also allows the designer to obtain what he or she must have.

It is necessary to evaluate negotiating goals and strategies before conducting any negotiations. The designer should write down what he or she must have and what can be conceded or modified. The designer should try to imagine how the shape of the contract will affect the future business relationship with the other party. Will it probably lead to success for both sides and more business, or will it fail to achieve what one side or the other desires?

When negotiating, the designer should keep written notes close at hand as to goals and strategies. Notes should be kept on the negotiations, too, since many conversations may be necessary before final agreement is reached. At certain points, the designer should compare where the negotiations have gone with the original goals. This will help evaluate whether the designer is conducting the negotiations according to plan.

Most negotiations are done over the telephone. The designer should decide when he or she wants to speak with the other party. Before calling, the designer should review the notes and be familiar with the points to be negotiated. If the designer wants the other party to call, the file should be kept close at hand, so there is no question as to where the negotiations stand when the call comes. If the designer is unprepared to negotiate when the other side calls, the only wise course is to call back. Negotiation demands the fullest attention and complete readiness on the part of the designer.

Oral Contracts

Despite all the forms in this book being written, it is worth addressing the question of oral contracts. There are certain contracts that must be written, such as a contract for services that will take more than one year to perform or, in many cases, a contract for the sale of goods worth more than $500. So, without delving into the full complexity of this subject, certain contracts can be oral. If the designer is faced with a party who has breached an oral contract, an attorney should certainly be consulted for advice. The designer should not give up simply because the contract was oral.

However, while some oral contracts are valid, a written contract is always best. Even people with the most scrupulous intentions do not always remember exactly what was said or whether a particular point was covered. Disputes and litigation are far more likely when a contract is oral rather than written. That is another reason to make the use of written forms, like those in this book, an integral part of the business practices of any designer whose work may someday have value.

Letter Contracts

If the designer feels sending a well-drafted form will be daunting to the other party, it is always possible to adopt the more informal approach of a letter that is signed by both parties. In this case, the contracts in this book will serve as valuable checklists for the content and negotiation of the letter contract. The last paragraph of the letter would say, "If the foregoing meets with your approval, please sign both copies of this letter beneath the words AGREED TO to make this a binding contract between us." At the bottom of the letter would be the words AGREED TO with the name of the other party, so he or she can sign. Again, if the other party is a company, the company name would be placed beneath the words AGREED TO, as well as the name of the individual who will sign and that individual's title. This would appear as follows:

AGREED TO:

XYZ Corporation

By: _____

Alice Hall, Vice President

Two copies of this letter are sent to the other party, who is instructed to sign both copies and return one copy to the designer for his or her files. To be cautious, the designer can send the letters unsigned and ask the other party to sign and return both copies, at which time the designer will sign and return one copy to the other party. This gives the other party an opportunity to review the final draft, but avoids a situation in which the other party might choose to delay signing, and the designer would not be able to offer a similar contract to another party because the first contract might still be signed.

If the designer should ever sign a contract that the other party does not sign and return, it should be remembered that any offer to enter into a contract can always be revoked up until the time that the contract is actually entered into. The designer can protect his or her position by being the one who is last to sign, by insisting that both parties meet to sign, or by stating in the letter a deadline by which the other party must sign.

Standard Provisions

The contracts in this book contain a number of standard provisions, called "boilerplate" by lawyers. These provisions are important, although they will not seem as exciting as the provisions that relate more directly to the designer and the design process. Since these provisions can be used in almost every contract and appear in a number of the contracts in this book, an explanation of each of the provisions is given here.

Amendment. Any amendment of this Agreement must be in writing and signed by both parties.

This guarantees that any changes the parties want will be made in writing. It avoids the possibility of one party relying on oral changes to the agreement. Courts, by the way, will rarely change a written contract based on testimony that there was an oral amendment of the contract.

Arbitration. All disputes arising under this Agreement shall be submitted to binding arbitration before _____ in the following location _____ and shall be settled in accordance with the rules of the American Arbitration Association. Judgment upon the arbitration award may be entered in any court having jurisdiction thereof. Notwithstanding the foregoing, either party may refuse to arbitrate when the dispute is for a sum of less than $_____.

Arbitration can offer a quicker and less expensive way to settle disputes than litigation. However, the designer would be wise to consult a local attorney and make sure this is wise in the jurisdiction where the lawsuit would be likely to take place. The arbitrator could be the American Arbitration Association or some other person or group that both parties trust. The designer would also want the arbitration to take place where he or she is located. If small claims court is easy to use in the jurisdiction where the designer would have to sue, it might be best to have the right not to arbitrate if the disputed amount is small enough to be brought into the small claims court. In this case, the designer would put the maximum amount that can be sued for in small claims court in the space at the end of the paragraph.

Assignment. This Agreement shall not be assigned by either party hereto, provided that the Designer shall have the right to assign monies due to the Designer hereunder.

By not allowing the assignment of a contract, both parties remain more certain with whom they are dealing. Of course, a company may be purchased by new owners. If the designer only wanted to do business with the people who owned the company when the contract was entered into, change of ownership might be stated as a ground for termination in the contract. On the other hand, money is impersonal, and there is no reason why the designer should not be able to assign the right to receive money.

Bankruptcy or Insolvency. If the Client shall become insolvent or if a petition in bankruptcy is filed against the Client, or a Receiver or Trustee is appointed for any of the Client's assets or property, or if a lien or attachment is obtained against any of the Client's assets, this Agreement shall immediately terminate, and the Client shall return to the Designer all of the Designer's work which is in the Client's possession and grant, convey, and transfer all rights in the work back to the Designer.

This provision seeks to lessen the impact on the designer of a client's bankruptcy. Such a provision could appear in a contract with corporate clients, merchandisers of products licensed by the designer, fabricators, or other suppliers. However, the bankruptcy law may impede the provision's effectiveness.

Complete Understanding. This Agreement constitutes the entire and complete understanding between the parties hereto, and no obligation, undertaking, warranty, representation, or covenant of any kind or nature has been made by either party to the other to induce the making of this Agreement, except as is expressly set forth herein.

This provision is intended to prevent either party from later claiming that any promises or obligations exist except those shown in the written contract. A shorter way to say this is, "This Agreement constitutes the entire understanding between the parties hereto."

Cumulative Rights. All rights, remedies, obligations, undertakings, warranties, representations, and covenants contained herein shall be cumulative, and none of them shall be in limitation of any other right, remedy, obligation, undertaking, warranty, representation, or covenant of either party.

This means that a benefit or obligation under one provision will not be made less because of a different benefit or obligation under another provision of the contract.

Death or Disability. In the event of the Designer's death or an incapacity of the Designer, making completion of the work impossible, this Agreement shall terminate.

A provision of this kind leaves a great deal to be determined. Will payments already made be kept by the designer or the designer's estate? And who will own the plans in whatever stage of completion has been reached? These issues are best resolved when the contract is negotiated.

Force Majeure. If either party hereto is unable to perform any of its obligations hereunder by reason of fire or other casualty, strike, act or order of a public authority, act of God, or other cause beyond the control of such party, then such party shall be excused from such performance during the pendency of such cause. In the event such inability to perform shall continue longer than _____ days, either party may terminate this Agreement by giving written notice to the other party.

This provision covers events beyond the control of the parties, such as a tidal wave or a war. Certainly, the time to perform the contract should be extended in such an event. There may be an issue as to how long an extension will be allowed. Also, if work has commenced and some payments have been made, the contract should cover what happens in the event of termination. For example, must payments be returned?

Governing Law. This Agreement shall be governed by the laws of the State of _____.

Usually, the designer would want the laws of his or her own state to govern the agreement.

Liquidated Damages. In the event of the failure of XYZ Corporation to deliver by the due date, the agreed-upon damages shall be $_____ for each day after the due date until delivery takes place, provided the amount of damages shall not exceed $_____.

Liquidated damages are an attempt to anticipate in the contract what damages will be caused by a breach of the contract. Such liquidated damages must be reasonable. If they are not, they will be considered a penalty and unenforceable.

Modification. This Agreement cannot be changed, modified, or discharged, in whole or in part, except by an instrument in writing, signed by the party against whom enforcement of any change, modification, or discharge is sought.

This requires that a change in the contract must at least be written and signed by the party against whom the change will be enforced. This provision should be compared to that for amendments, which requires any modification to be in writing and signed by both parties. At the least, however, this provision explicitly avoids a claim that an oral modification has been made of a written contract. Courts will almost invariably give greater weight to a written document than to testimony about oral agreements.

Notices and Changes of Address. All notices shall be sent to the Designer at the following address: _____ and to the Purchaser at the following address: _____. Each party shall be given written notification of any change of address prior to the date of said change.

Contracts often require the giving of notice. This provision facilitates giving notice by providing correct addresses and requiring notification of any change of address.

Successors and Assigns. This Agreement shall be binding upon and inure to the benefit of the parties hereto and their respective heirs, executors, administrators, successors, and assigns.

This makes the contract binding on anyone who takes the place of one of the parties, whether due to death or simply an assignment of the contract. With commissioned works, death or disability of the designer can raise complex questions about completion and ownership of the art. The issues must be resolved in the contract. Note, the standard provision on assignment, in fact, does not allow assignment, but that provision could always be modified in the original contract or by a later written, signed amendment to the contract.

Time. Time is of the essence.

This requires each party to perform exactly to whatever time commitments they have made or be in breach of the contract. It is not a wise provision for the designer to agree to, since being a few days late in performance could cause the loss of all benefits under the contract.

Waivers. No waiver by either party of any of the terms or conditions of this Agreement shall be deemed or construed to be a waiver of such term or condition for the future, or of any subsequent breach thereof.

This means that if one party waives a right under the contract, that party has not waived the right forever and can demand that the other party perform at the next opportunity. So, the designer who allowed a client not to pay on time would still have the right to demand payment. And if the client breached the contract in some other way, such as not returning a model for design, the fact that the designer allowed this once would not prevent the designer from suing for such a breach in the future.

Warranty and Indemnity. The Designer hereby warrants that he or she is the sole creator of the Work and owns all rights granted under this Agreement. The Designer agrees to indemnify and hold harmless the Client from any and all claims, demands, payments, expenses, legal fees, or other costs based on an actual breach of the foregoing warranties.

This provision protects one party against damaging actions that may have been taken by the other party. Often, one party will warrant that something is true and then indemnify and hold the other party harmless in the event that it is not true. For example, a designer selling a design may be asked to warrant that the design is not plagiarized. Or the designer may ask a supplier to warrant this to the designer. If, in fact, the design has been plagiarized, this would breach the warranty. The party breaching the warranty would be obligated to protect the other party who has been injured by the warranty not being true.

Using the Checklists

Having reviewed the basics of dealing with the business and legal forms, it is time to move on to the checklists that will make the forms most useful.

These checklists focus on the key points to be observed when using the forms. On the organizational forms, the boxes can be checked when the different aspects of the use of the form have been considered. For the contracts, the checklists cover all the points that may be negotiated, whether or not they are in the contract. When, in fact, a point is covered in the contract already, the appropriate paragraph is indicated in the checklist. These checklists are especially valuable to use when reviewing a contract offered to the designer by someone else.

For the contracts, if the designer is providing the form, the boxes can be checked to be certain all the important points are covered. If the designer is reviewing someone else's form, checking the boxes will show which points they have covered and which points may have to be altered or added. By using the paragraph numbers in the checklist, the other party's provision can be quickly compared with a provision that would favor the designer, as presented in this book. Each checklist for a contract concludes with the suggestion that the standard provisions be reviewed to see if any should be added to what the form provides. Of course, the designer does not have to include every point on the checklist in a contract, but being aware of these points will be helpful.

Job Index

The job index is the source of all project numbers. The purpose of the project number is to identify all the components of an assignment. Job numbers are used on job sheets, time sheets, purchase orders, invoices (incoming and outgoing), transmittals, and all other project records. It is the link to all project-related costs and materials.

As soon as a prospective assignment is introduced, it is useful to open a job number to begin tracking time and costs, including the effort it takes to take the project from prospect to reality. In other words, even if the job does not proceed past the proposal stage, it is helpful to look back at the end of the year and calculate how many jobs were pursued and how many became active assignments. Additionally, while expenditures toward an unsuccessful bid may not be reimbursable, these expenditures can be tax deductible as part of the firm's marketing costs. Some firms choose to use separate job numbers for projects that are in the marketing stage from those that are in production. The index can be organized in a number of ways: chronological, alphabetical by client, or by project type. Regardless of the method used to classify projects, this comprehensive index of projects is the simplest and clearest way of identifying all past and current projects, as well as who is leading them and how long they have been active, or when they were completed.

The "project number" index, or list, can begin with any number, but then should follow consecutively thereafter. The numbers may simply be keyed to start dates, such as "2005–52," or in the year 2005, job number 52. Aside from how the numerical order is determined, it is best to keep the number uncomplicated and as brief as possible since it will be used frequently. Always keep the numbers in sequence. Do not fill in the numbers in advance since some assignments may need additional space. For jobs with several distinct subparts that need to be tracked

separately for billing purposes, use one project number for the overall name of the assignment/client and use subnumbers or subletters to indicate the different parts of the job (also keep separate job sheets for such subnumbers or letters). See example A.

Another way to establish job numbers is to use a code. Example A is based on the system shown in Form 1, whereby in PA05-01, "PA" stands for the first two letters of the client company name, "05" stands for the year of project inception, and "01" indicates it was the first job that year. "A," "B," and "C" refer to the individual components of the project. If jobs are unrelated, but have the same client and same start date, use separate project numbers, as shown in example B.

Filling in the Form

In the "date opened" column, fill in the date the project is assigned to the firm. It may be the date of the signed project agreement, or it may be the date of the first client briefing, if your practice is to track pre-contract expenses. "Date closed" is the date that the final payment was received. In the "client" column, fill in the client's name—regardless of whether it is an individual or a company or some other entity. Under "job name," identify the job with a name that is clearly distinguished from other projects in the office.

Under "project lead," fill in the name of the lead designer or project manager who will have primary responsibility for the assignment. Then fill in the job number.

Using the Form

Use project numbers on all project-related forms, drawings, transmittals, and other correspondence. It is the central link to all information about a project.

Example A

Date Opened	Date Closed	Client	Job Name	Project Lead	Project #
7/1/05		Parks Group	Street Furnishings	Natalie Block	PA05-1
"	9/1/06	" "	-Bench	Spencer Boon	PA05-1A
"		" "	-Kiosk	Spencer Boon	PA05-1B
"		" "	-Street Lamp	Clara Danom	PA05-1C
8/19/05		Arc Company	Window Fan	Stuart Brand	AR05-2

Example B

Date Opened	Date Closed	Client	Job Name	Project Lead	Project #
3/9/06		Applegate Inc.	Laptop Tray	Zoltan Dean	AP05-11
"		" "	Floor Lamp	Clara Danom	AP05-12
"		" "	Task Chair	Zoltan Dean	AP05-13

Post the job index near special printers, the messenger desk, and other areas where employees can easily refer to it and be encouraged to use the job numbers consistently. It greatly simplifies record keeping.

Open job files, both computer and physical. Some firms use large three-ring binders to keep all hard copy documents organized. Label the files and project folders or envelopes with the job number, client and job name, as well as the project start date.

Job files (digital and hard copy) are used to store copies of agreements, work change orders, receipts, and all other project materials. It is useful to separate correspondence and other nonfinancial documents from materials that will serve as backup for billing purposes. (The most popular method of filing documents is in chronological order, with the most recent item on top.) If you know that you will be required to show backup for reimbursable expenses, a separate folder containing these receipts is a convenient place to store these copies.

When the project is completed and billed, simply store its files and folders in chronological/alphabetical order for future reference. Archive all digital documents to CD-ROM or DVD disks, or some other backup media; make sure to print out the file directory, to make it easier to find specific documents. Quite often it is helpful to be able to check back to earlier proposals and contracts when estimating or negotiating or planning new projects.

Job Index

Date Opened	Date Closed	Client	Job Name	Project Lead	Job Number

Job Sheet

Cost accounting, or keeping track of time and costs expended on a project, is an essential indicator of whether a project is profitable or not, as well as useful in maintaining regular and accurate billings. The job sheet is the detailed record of all time and costs incurred during the course of a project. It is useful to record time and costs regardless of the fee arrangement since even miscellaneous unbillable time and expenses have an impact on profits when all real costs are fully known. Items such as special materials, extended research time, and very long and frequent client meetings not originally calculated into the fee may significantly diminish what seemed to be an acceptably profitable job. The advantage of knowing all costs, including extras, is the possibility of more astutely negotiating time and fees of prospective assignments, or perhaps renegotiating a current one.

Job sheets can be used to analyze the following information:

Time. In addition to noting the time spent on the whole project, this category can be further detailed to indicate specific aspects or phases of work. For example, in the design of a coffee maker: it may be necessary or useful to know how much time was spent on the internal mechanisms, versus the exterior housing; or how much time was spent generating design drawings separate from testing, and so on.

Billable Costs. Depending upon the fee arrangement, industrial designers might charge back the cost of certain materials and/or outsourced services, such as trademark searches, model making, and other expenses such as travel, messengers, and conferencing telecommunications. All of these may be billable depending upon the contract terms. Non-billable items can be noted with an "NB" to the right of the "total" column.

Profitability. Accurate records are indispensable in determining whether a fee is adequate for the time and expenses dedicated to a project. Here are three calculations to examine profitability:

1. To arrive at the Gross Profit, subtract total actual (non-reimbursable) project costs from the total fee amount billed (not including taxes):

   Fee minus Non-reimbursable Project Costs* = Gross Profit

   $100,000 minus $15,000 = $85,000

   *These would be absorbed costs, not included in the fee nor charged back to the client, e.g., consultants with special skill sets who are not on staff, but who perhaps would be at another firm, so the client will not pay extra for their expertise.

2. To calculate the Net Profit, subtract the billable cost of all project personnel (including an overhead calculation) from the Gross Profit:

   Gross Profit minus Billable Project Labor* = Net Profit

   $85,000 minus $65,000 = $20,000

   *Billable Project Labor is the marked up or billing rate of every person who works on the project—and it must include all overhead costs. Overhead includes all non-project specific costs that are involved in running your business. These items include, but are not limited to: rent, insurance, utilities, maintenance, general supplies, postage and subscriptions, bookkeeping and legal services, computers, leases, company advertising/marketing, salaries for support personnel, all salary-related benefits, and so on. While none of these items are billable to specific projects, they, along with a profit margin and contingency allowance, must be figured into the fee structure of every assignment.

The overhead factor is most usually added to the hourly rates of personnel dedicated to a project. To calculate overhead expenses, add up all of the monthly business costs (which are not billable as reimbursable expenses), multiply by twelve to get the annual number and divide this total by the number of working hours available per staff member in the year; the resulting number is the dollar amount you have to add to each staff person's basic hourly rate. For example (calculations are rounded):

Rent	$2,000
Utilities	$250
Leases	$250
Supplies	$250
Insurance	$200
Legal	$200
Advertising	$100
Marketing	$100
P/T Bookkeeper	$700
Misc.	$200

Total Monthly Overhead: $4,250

Total Annual Overhead Expenses: $4,250 × 12 = **$51,000**

Number of billable employees: 4

Number of billable hours in a year per person: 1,680 hours (48 weeks × 35 hrs/wk)*

Number of billable hours in a year: 6,720 hours (4 people × 1,680 hours)

Overhead to be added to each billable hour per person: $7.60 ($51,000/6,720 hours)

*Logic and experience will show that 100 percent of billable time is not typical of most employees, particularly high-level managers who spend considerable time on marketing, administration, and personnel issues. Be realistic in calculating billable time.

So, if you have a designer who is earning $60,000 per year, add approximately 38 percent to cover benefits (including paid time out, health plan, etc.), social security and taxes, for a total cost to the business of $82,800 per year; this person's billing rate, before including overhead, is $49.30 per hour ($82,800/1,680); add the overhead, and the billing rate is $56.90 ($49.30 + $7.60). With a profit target of 20 percent, and 5 percent for contingencies, the final billing rate for this individual is (rounded up) $72.00 per hour ($56.90 × 1.20 = $68.28; $68.28 × 1.05 = $71.69).

Net Profit divided by Fee = Percent Profit Margin

$20,000 divided by $100,000 = 20 percent

Calculating profit into fees and hourly rates is not an exact science. Generally designers try to reach for a minimum profit of 20 percent.

Some companies simply multiply an employee's base hourly salary by a factor of 3, or 4, or 5 (if the market can bear it) to cover salaries, benefits and other employee costs, overhead, profit, and contingencies. Firms often assign tiers of billing rates for different kinds of personnel—senior personnel are all billed at one rate amount, mid-level at another, and junior level at still another. This simplifies billing and also reduces the chance of staff members discovering each other's salaries.

Assuming this firm has four full-time staff members, including the principal, the overhead outlined above, and a profit margin goal of 25 percent, how much revenue will the firm have to generate in a month to cover salaries, employee benefits, overhead, and profit? (For the purposes of this exercise, we are leaving out federal, state, and local tax calculations).

Using this calculation for overhead expenses, this business must produce $501,638 of gross revenue for the year, not including those project expenses that are reimbursable (while you have to pay those expenses up front, you will be reimbursed, hopefully with a markup). In order for this firm to stay in business and make its target profit, it will need either one large project at this fee, or a series of different projects at varying fees that add up to at least $501,638 for the year.

Salaries	$240,000	Principal @ $100,000; 1 designer @ $60,000; 2 junior designers @ $40,000 each
Salary-related expenses	$ 91,200	38 percent of salaries
Overhead	$ 51,000	See calculations at left for overhead
Profit goal	**$ 95,550**	**25 percent of salaries + expenses + overhead**
Contingencies	**$ 23,888**	**5 percent of salaries + expenses + overhead + profit**
Total	$501,638	

An ideal strategy is to reduce as much as possible the non-reimbursable items on every project. That is, make sure that all time and costs are fully reimbursable. This tactic points to the need to develop extremely accurate estimates for the time and materials that every project is expected to consume. Every project has glitches and every business experiences unexpected expenses, such as increases in the cost of a basic service, the cost of a tax audit, or the sudden need to replace some indispensable equipment.

The contingencies allocation, shown here as 5 percent, is the minimal amount of protection, or cushion, any business (and project) needs to cover unavoidable, unanticipated, non-reimbursable costs that are not covered by the overhead allocation.

Naturally, hourly rates have to have a reasonable relation to what the competition is charging for similar work and what the market will bear. Flexibility is important in determining fees and hourly rates—but you must never lose sight of the need to cover the ongoing expenses of your business. To cover slow periods, a business owner must either have a reasonable amount of savings or a credit line to cover overhead expenses while searching for new assignments. The ability to adjust expenses quickly in response to changes in the volume of work is also helpful, but so is a healthy amount of surplus savings. It is also recommended that no matter how busy you are with ongoing projects, you should never stop marketing for future ones.

Project Types
In general, high-volume, mass-market commercial products command higher design and development fees than small-run, highly crafted products. On the other hand, the requirements of commercial projects are usually more complex and take longer to complete, increasing the chances of unanticipated problems and cost overruns. However, commercial product development may be more predictable than highly customized products that involve the subjective nature of client tastes and their abilities to make firm decisions. The more experience an industrial design firm has in niche areas, whether it's product type, industry type, technology type, or some other type of category, the greater its expertise in estimating costs, fees, and profits in these areas.

Filling in the Form
Enter the job number, its name and the client contact information indicated. Note different billing and shipping addresses, if necessary. Referral source information is useful in determining the effectiveness of your marketing or sales efforts. If the referral was a former client, it should signal a valuable source of future references, as well as a special personal acknowledgment to that client. Project team information is important to note in that it is a quick reference for the person preparing periodic cost accounts and billing.

The date information section is useful for cost accounting in several ways. First, if a project is billable at specific stages or completion dates, then it is necessary to keep track of these dates. Second, since project deliverables are predicated upon specific approvals, it is important to be able to show the client a record of slipping sign-off dates that affect deliverable dates. Additionally, a clear record of dates is very useful in understanding those areas of the project that were particularly efficient or inefficient, and applying this knowledge to future project plans. The notes area is helpful to record information that requires explanation.

The billing area of this form is the place to note the terms of the fees to be charged, whether fee plus costs, time plus costs, and other specific details of the fee agreement. Keep a record of billings under "invoice date, invoice number, and invoice amount" for a quick review of the project's billing history. It is handy to have all billing information available on one sheet. The work change orders section is particularly important to keep up to date. It is a summary of the work changes approved by the client and spells out the amounts of additional (or reduced) billing as well as changes in deliverable dates. A quick summary on this sheet will eliminate time wasted looking through stacks of paperwork.

Using the Form

Start a job sheet as soon as a job number has been assigned to the project from the job index, which can either be a worksheet in your electronic project file or in front of the job sheet binder. Posting should take place after time sheets have been collected and approved for payroll and when job-related bills are being paid. Indicating job numbers on time sheets and payable invoices makes it easier to post these items to individual job sheets.

Messenger, courier, and other service logs should have a jobs index nearby for easy reference to job numbers. When you receive bills for these services, it will be easy to reference these expenses to projects. When a time plus cost-based project is ready for billing, tally all related items separately. For example, add messenger, material costs, parts, supplies travel, and so on, separately. Add up project personnel time and/or total cost (if freelancer

or contractor) separately. For billing, either itemize each cost type separately or combine under general headings by project areas or tasks. Remember to factor markups as appropriate. You may or may not choose to show these calculations on your invoices, depending on the agreement you have with your client.

Filling in the Form (Summary of Costs)

The "summary of costs" is a comprehensive view of the project's estimated and actual job costs with labor and expenses shown separately. You may wish to simply total both labor and expenses for each line item. Use this summary as a worksheet when calculating the total costs of the completed project. You can add additional "actual" columns for interim summaries for partial billing—but do include a final total column.

Filling in the Form (Cost Account)

Uncomplicated short-term assignments may only need a couple of these supplementary pages. When projects are complex or continue for long periods of time, there may be many supplementary pages to list the on-going time and costs. This form allows for the listing of billables, line by line. The date that the item is posted on this sheet goes in the "date" column. "Item" is the name of either the person (staff or freelance) or the vendor/supplier. Under "description," fill in the invoice number and date; in the case of a staff person, fill in the date of the timesheet or the week-ending date; put the number of billable hours under "hrs/rate"; there is no need to show their rates here, since you have already included that in the project team section at beginning of the form. For freelancers, you can either indicate their hours and rates in this column, or put their invoice fee under "total."

The "expense" column is for invoices received that separate labor from expenses. For example, a rendering artist might submit an invoice like this:

Presentation drawings:

18 hours @ $40/hour	$720

Expenses:

Library Reference Room Fees:	$12
Books:	$75
Transportation (NYC/Boston/NYC):	$195
Subtotal:	$282
Total:	$1,002.00

For jobs that require periodic billing for fees and costs, draw a bold line or skip a space under the last item included in each separate invoice and jot down the invoice number by this line so that it will be easy to see where to start tallying for the next invoice.

Job Sheet

JOB

Number _____

Name _____

CLIENT

Name _____

Address _____

Phone _____

Cell _____

Fax _____

Email _____

BILLING INFORMATION

Name _____

Address _____

Phone/Fax _____

SHIPPING INFORMATION

Name _____

Address _____

Phone/Fax _____

INFORMATION

Date _____

Referral/Source _____

Name _____

Address _____

Phone _____

Sales/Marketing _____

P.O. # _____

PROJECT TEAM

Title	Name	Billing Rate	Notes

DATES

	Target Date	Actual Date	Notes
Contract Start			
Intake Meeting(s)			
Strategic Analysis and Planning			
Start			
Final Presentation			
Sign-Off			
Creative Development			
Start			
Final Presentation			
Sign-Off			
Implementation			
Start			
Final Sign-Off			
Launch			

BILLING

Fee Information   **Billing Schedule**   **Notes**

Fee Amount
Time & Materials
Other
Total

Invoice Date	Invoice Number	Billed To	Fee	OOPs	Total

WORK CHANGE ORDERS

Date	Order Number	Item	Hrs/Rate	Expenses	Total	Notes

COST ACCOUNT

Date	Item	Description	Hrs	Rate	Expenses	Total	Billable/Non	Notes
Subtotal	January							
Subtotal	February							
Subtotal	March							

SUMMARY OF COSTS

Item	Est. Time	Est. Cost of Labor	Est. Expenses	Est. Total	Actual Time	Actual Cost of Labor	Actual Expenses	Actual Total
By Phase								
Strategic Anaylsis and Planning								
Creative Development								
Implementation								
Total								
By Detail								
Department								
Executive Management								
R&D								
Design								
Engineering								
Technology								
Manufacturing								
Quality Control								
Marketing								
Packaging								
Sales								
Legal								
Finance								
H.R.								
Etc.								
Total								
Task								
Pre-Planning								
Decision Screen								
Needs Analysis								
Audits								
Segmentation								
Ideation								
Evaluation								
Design Development								
Testing								
Pilot Production								
Etc.								
Total								
Expense Type								
Color Copies								
Color Transfers/Other Color Processes								
Computer Prints								
Courier Services/ Shipping								
Fabrication								

Item	Est. Time	Est. Cost of Labor	Est. Expenses	Est. Total	Actual Time	Actual Cost of Labor	Actual Expenses	Actual Total
Illustration/Other Rendering								
Legal Fees								
Local Messengers								
Modelmaking								
Photographic & A/V Supplies								
Photographic Prints								
Photography								
Portable Media (Disks, CD's, etc.)								
Presentation Materials								
Prototypes								
Reference Materials								
Reproduction								
Research & Database Services								
Research Materials								
Scanning								
Shipping								
Special Supplies/Materials								
Telecommunications								
Travel								
Airfare								
Ground Transportation								
Lodging								
Meals								
Entertainment								
Other								
Total								

Comprehensive Schedule

Industrial design is a particularly deadline-driven profession. There is tremendous pressure to launch products on time for a number of reasons. One is to have early advantage of new product features before competitors have a chance to copy or improve them. Another is that delays in design and/or production add extra cost to a product's development. Additionally, there are usually marketing efforts, advertising, and distribution chain issues that require thoughtful planning and a keen understanding of logistics.

Project leaders/managers are responsible for coordinating the activities of their staff and of numerous other internal and external audiences. They deal with the delivery of technology, parts, and materials in order to complete assignments on time. In order to calculate the blocks of time needed to accomplish different tasks and make sure that the right people and resources are in place to meet deadlines, the primary project leader should work out an activity schedule as soon possible for the overall project; each division or department leader in the different areas of the company should work out their group's production and resource schedule. The best time to develop the production schedule is after there is certain knowledge of the details of the product and prior to, or during, project estimating.

A common practice is to calculate schedules by working backward from the final due date for project completion. If this process reveals that there is not enough time or resources to accomplish the task, it's still early enough to effect changes in the scope of the assignment.

Two scheduling forms are offered in this section. The first gives a snapshot view of all the projects in the company, or the individual departments, such as the design studio. It is meant to highlight top-level information,

such as the timing of project phases, major event dates, personnel and special notes. Its particular usefulness is that it immediately reveals overlaps in the use of staff and other resources, particularly for important meetings, site or delivery supervision, and other critical events.

The second form is designed for individual project use. It lists the major phases of a project, and within each phase the range of items that may be required. Please note that if your projects are less extensive (or more) that you may easily remove (or add) items that are relevant to your projects.

Filling in the Form (Form One— All Projects)

List all the active jobs in the company, department, or studio. You may wish to include those that are being marketed, but not yet contracted, in order to anticipate possible conflicts in staff scheduling and thereby allow for adjusting start dates on new projects. List the project phases, their start and end dates, and the key due dates for each phase. List the names of the staff members working within each phase, as well as the consultants and the vendors (under "sources"). In the "notes" area, highlight any items that need special attention.

Using the Form

Keep the "all projects" schedule up to date as project phases shift, and remove old information. This form is not a record of past events; it is a living document that needs continual updating. It is the best way to prevent double-booking important meetings and avoid conflicts in scheduling staff, freelance and other resources.

Industrial design is not a linear activity—phases and activities will overlap, particularly during implementation.

Charts and graphs are a simple visual aid to help track schedules. It is relatively easy to convert data from spreadsheet format to charts and graphs in Microsoft Excel. Once you are familiar with this feature, track staff overlaps by using the bar graph feature. A more sophisticated visual tracking system is available in Microsoft Project. The application is not particularly intuitive, but once you become familiar with its features (and quirks), it does have several useful features. First of all, it presents scheduling data in Gantt and PERT chart formats. The Gantt chart presents your list of activities along with the scheduled start and completion dates via a horizontal bar chart. It also includes resource usage whereby you list the personnel required for each activity, their rates, and the percentage of time they will spend on each activity. It offers a wide variety of report views, some of which are standard, some of which can be customized. A calendar program may also serve to highlight project phases, due dates and the use of personnel, but generally, there is limited space to provide all of this information at one time.

Filling in the Form (Form Two—Individual Projects)

Fill in the "job number" and "job name" areas. List start and end dates under "schedule." Name the person who is responsible for providing the key deliverables ("phase lead"). Insert the actual start and end dates in order to compare against scheduled dates.

Using the Form

It is very important to either have experience or else thoroughly research the amount of lead time needed to build models and prototypes, develop or adapt technologies, obtain unique parts and materials, conduct testing, and so on. Attention must be paid to lead time since availability, development, and manufacturing of key elements—even shipping times may vary. While it is possible to rush orders, the added cost in time and money can be excessive.

The individual projects production schedule can be shared with the client, supplementing the status report for client (Form 9). It provides baseline scheduling information that can be referenced when the client suggests changes that affect delivery dates.

Comprehensive Schedule

ALL PROJECTS

Job Number	Job Name	Status - Current Phase/Activity	Next Steps	Project Lead	Notes

INDIVIDUAL PROJECTS

Date	Job Number	Job Name	Project Lead	Schedule		Actual	
Phase	**Item**	**Notes**	**Phase Lead**	**Start**	**End**	**Start**	**End**
Strategic Analysis and Planning							
	Project Commencement/ Pre-Planning						
	Decision Screen						
	Needs Analysis						
	Intake Meetings						
	Audits						
	Client and/or Industry						
	Market						
	Product						
	Develop Targeted Strategy						
	Team Selection						
	Product Brief						
	Design Brief						
	Budget/Schedule						
	Presentation						
	Revisions						
	Sign-Off						
Creative Development							
	Concept Development/ Ideation/Brainstorming						
	Renderings						
	Model-making						
	Evaluation						
	Strategic						
	Functional						
	Refinement						
	Sign-Off						
	Design Development						
	Prototyping						
	Testing						
	Refinement						
	Testing						
	Refinement						
	Testing						
	Refinement						
	Specifications						
	Presentation/Sample Models & Boards						

Date	Job Number	Job Name	Project Lead	Schedule		Actual	
Phase	Item	Notes	Phase Lead	Start	End	Start	End
	Design Development						
	Presentation						
	Revisions						
	Budget/Schedule						
	(Re)Estimates						
	Sign-Off						
	Pilot Production						
	Blueprints and Specs						
	Sub-Assembly						
	Testing						
	Refinement						
	Testing						
	Refinement						
	Testing						
	Refinement						
	Contract Documents						
	and Bids						
	Bid Drawings and Specs						
	Bid Review						
	Client Review of Bids						
	Revisions						
	Vendor Selection						
	Contract Documents						
	Sign-Off						
Implementation							
	Manufacturing Plan						
	Location/Space/Layout						
	Equipment						
	Tools						
	Technology						
	Methodology						
	Production Development						
	Personnel-Recruitment						
	Personnel-Training						
	Sourcing Parts and						
	Materials						
	Trial Production						
	Testing						
	Refinement						
	Testing						
	Refinement						
	Sign-Off						
	Production Ramp-up						
	Troubleshooting						
	Quality Control						
Launch							
	Packaging						
	Marketing						
	Advertising						
	Launch Events						
	Distribution						
	Customer Support						

Project Plan and Budget Estimate

There are a few of ways to charge for the cost of industrial design endeavors. Methods include a flat fee; for consultants, hourly rates plus expenses; and for those partnering in the ownership of the product, a percentage of revenue from products sold (royalty). Depending upon circumstances, one or a combination of these methods can apply.

This preliminary plan and budget form is intended to help the designer accurately calculate every aspect of a proposed project as well as serve as a comprehensive checklist of items to be considered. It is meant to be particularly useful in cases where the designer does not have a great deal of experience in the proposed project type or when detailed records of past projects are unavailable. With the results derived from these calculations, the designer/entrepreneur should be able to better anticipate the actual cost of producing the job.

Assuming that you are basing the project fee on a time and materials basis, here is a form that lends itself to many purposes. You can use it internally to calculate all project expenses by labor and materials costs, phase by phase. For client presentations, you can show estimated time and costs by line item, but hide the "billing rate" rows. As a budget and schedule status review, it serves to summarize periodically actual versus budgeted costs and schedule.

Filling in the Form

While this form is intended to be as complete as possible, it is better to use only those line items that apply to the project at hand to avoid confusion. With the inclusion or deletion of specific column headings, this form may start out as the initial estimating spreadsheet used to determine detailed costs. Used as an estimating worksheet, it is most important to break down every aspect of the project into its separate components in order to determine how the complexity and scale of the job will determine the final costs.

Depending upon its intended use, check the appropriate box for the preferred title of the form. As either a "preliminary budget and schedule" or as a "budget and schedule status review," this form may be an internal document only; or it may be modified to show to clients.

Fill in the client information, including name, addresses, and phone numbers. Include the project name and a brief description. In the event you have already assigned this project a number, fill it in. Most of this information is also found on the job information sheet (which is not usually seen by the design staff); it is easy to simply copy and paste the areas, as appropriate.

Two options are offered here. One, is to estimate the project by the various competencies or departments that will be participating in its development, such as executive management, research and development, engineering, design, etc. The second is by phase of work, whereby the development of a project is organized by stages of development. The estimating format is the area for breaking out the detailed project plan. Check the "assumption," or, basis of rates, whether hourly, daily, or weekly. Regardless of which you select, make sure that all of the estimates are based on the selected measure of time. If you select anything other than an hourly rate, for example, daily or weekly, be careful to adjust for the percentage of time ("percent time") that individual will work on this project during the day or week. In some instances, designers may be involved with more than one project at the same time. So, if your lead designer is going to spend half his or her time on developing concept schematics for one week on this project, and you are basing this estimate on weekly rates, show that as .5 (50 percent) for that week.

On the "billing rate" line, fill in each person's rate, depending on whether you have selected to estimate this project on an hourly, daily or weekly rate. These rates should be the billing rates—including benefits, overhead, etc., as discussed for the job sheet (Form 2).

For every task listed under the main groupings of activities, fill in the estimated amount of time each person will spend working on the project. This form has preset formulas that will automatically feed the rate times the time, per person, into the column for "total labor" amount. If you choose to add or delete lines or columns, make sure to check and adjust the formulas, as necessary. This form does not contain any macros; it is formatted for simple calculations. The form also has built-in calculations to add subtotals and the final total is the sum of all the subtotals.

In the "materials" column, put in the total cost of materials for each of the tasks listed; under "vendors," indicate the anticipated fees and costs that vendors will charge. In the "cost" section of columns, the "budget" column has already been filled in by the automated formula. In the "to-date" column, put in the total costs actually incurred at set intervals. These intervals may be weekly, monthly, or by periodic billings. The "balance" column will automatically subtract the "to-date" amount from the "budget" amount to show how much money is left, or overspent.

The "schedule" column can show either the overall time allotted for each task, or it can track the amount of labor estimated and used. If it is going to be used to track labor time, a formula can be inserted that adds up all the time shown in the "labor" columns. If the "schedule" column is being used to show overall time, then that information will have to be inserted manually.

Using the Form

Microsoft Project, which is also discussed in the comprehensive schedule section (Form 3), is a very useful software program that helps construct project schedules and budgets. It lends itself well to design productions and is also widely available for either PC or Mac. While it requires a fair learning curve to master, and is somewhat rigid, it does provide a wide variety of reports that can be used for estimating time

and costs, tracking expenditures and schedules and client communications.

Considered one of the safest ways of billing projects, retainers are actually based on the hourly rate method and protect the designer from clients who may not make timely decisions or projects that are vaguely defined. It requires accurate tracking of all personnel working on the project (using timesheets) and the calculation of hourly rates that include not only the base salary and all fringe benefits per person, but also markups for the firm's overhead costs and desired profit margin. (Overhead costs, billing rates, and profit margins are discussed in greater detail in the job sheet section of this book.) Additionally, all project-related expenses are billed separately, with or without markups, depending upon the terms negotiated for the project. It is customary to provide clients with copies of invoices for such items. It is unnecessary to show backup for projects that are billed on an all-inclusive, flat fee basis.

Regardless of the type or scale of the product, accurate estimating is the most important factor in assuring profitability. The key to estimating accuracy is to understand exactly what is involved in the successful completion of the project, from beginning to end.

It is also important to have some familiarity with the client to be able to judge how much time it will take to turn around approvals, revisions, and final signoffs. For repeat clients, this is made evident by past project experiences documented by timesheets and job summaries. In the case of new and unfamiliar clients, a fair measure of the client may be based on the pre-design phase. Otherwise, you can build in a contingency factor (a percentage applied to either hourly rates or total costs) to cover client delays leading to schedule "creep" (the slipping and extension of scheduled deliverables leading to additional costs for the design firm).

Technology developers, electronic specialists, mechanical engineering, and other trades that may be involved in the design and production of a product as contributors/vendors, have their own, specific methodologies for estimating costs. It is very important to know the differences in work time of the different trades and/or departments that may be contributing to the effort. Allow adequate lead time for their necessary input and communicate frequently and in detail about the product at all stages of its development.

By hiding the columns used for calculating "estimate factors," this form can serve as a detailed estimate form that the client sees. In this instance, it is particularly useful to include a statement at the bottom of this form to ensure that the client recognizes that this is an estimate and not a contract.

Finally, as the job proceeds, it can be used either as the internal summary of ongoing costs versus budget, or as a status report to the client, showing costs-to-date versus budget on an item-by-item basis. As an internal document, use it to track how much time and money have been expended on specific job components and how much is left in the budget to complete the job. The routine monitoring of ongoing costs is very useful in detecting areas that are either significantly over or under budget, allowing for budgetary and scheduling adjustments.

There is no standard requirement or practice about what industrial designers report to clients. Many would prefer to simply bill out a project at regular intervals against the agreed fee, making adjustments only for work change orders (Form 10). Government and technology clients often have more stringent reporting requirements, particularly for large projects and most clients find it reassuring to receive status reports at regular intervals.

Project Plan and Budget Estimate

❑ Preliminary Budget & Schedule   ❑ Budget & Schedule Status Review

CLIENT
Name
Address
Phone
Cell
Fax
Other Contact Information

By Competency/Department   ASSUMPTION: RATES ARE PER   ❑ HOUR ❑ DAY ❑ WEEK

See Job Sheet for headcount, hours, and billing rate   **PROJECT TEAM**

	Name/Title	Name/Title	Name/Title	Name/Title
BILLING RATE	$	$	$	$
% TIME	0%	0%	0%	0%
Executive Management				
R&D				
Design				
Engineering				
Technology				
Manufacturing				
Quality Control				
Marketing				
Packaging				
Sales				
Legal				
Finance				
H.R.				
Etc.				
Total-Time				
Total-Cost	$	$	$	$

By Phase of Work   ASSUMPTION: TIMING AND RATES ARE PER   ❑ HOUR ❑ DAY ❑ WEEK

See Job Sheet for headcount, hours, and billing rate   **PROJECT TEAM**

Strategic Analysis and Planning	Name/Title	Name/Title	Name/Title	Name/Title
BILLING RATE	$	$	$	$
%TIME	0%	0%	0%	0%
Project Commencement/Pre-Planning				
Decision Screen				
Needs Analysis				
Intake Meetings				
Audits				
Client and/or Industry				
Market				
Product				
Targeted Strategy Development				

PROJECT _____   Date _____

Name _____   By _____

Number _____

Description _____

ESTIMATE FACTORS			COST			SCHEDULE		
Total Labor	Materials	Vendors	Budget	To-Date	Balance	Allocated	To-Date	Balance
$	$	$	$	$	$			

ESTIMATE FACTORS			COST			SCHEDULE		
Total Labor	Materials	Vendors	Budget	To-Date	Balance	Allocated	To-Date	Balance

	PROJECT TEAM			
Strategic Analysis and Planning	**Name/Title**	**Name/Title**	**Name/Title**	**Name/Title**
Team Selection				
Product Brief				
Design Brief				
Budget/Schedule				
Presentation				
Revisions				
Subtotal-Time				
Subtotal-Cost	$	$	$	$

Creative Development	**Name/Title**	**Name/Title**	**Name/Title**	**Name/Title**
BILLING RATE	$	$	$	$
% TIME	0%	0%	0%	0%
Ideation - Brainstorming/Concept				
Renderings				
Modelmaking				
Evaluation				
Strategic				
Functional				
Refinements/Revisions				
Prototyping				
Testing				
Refinements/Revisions				
Concepts Presentation				
Client Sign-Off				
Design Development				
Specifications				
Testing				
Refinements/Revisions				
Design Development Presentation				
Refinements/Revisions				
Client Sign-Off				
Pilot Production				
Sub-Assembly				
Testing				
Refinements/Revisions				
Final Specs and Blueprints				
Contract Documents and Bids				
Bid Drawings and Specs				
Bid Review				
Client Review of Bids				
Refinement/Revisions				
Client Sign-Off				
Vendor Selection & Contract Documents				
Budget/Schedule (Re)Estimates				
Subtotal-Time				
Subtotal-Cost	$	$	$	$

ESTIMATE FACTORS			COST			SCHEDULE		
Total Labor	Materials	Vendors	Budget	To-Date	Balance	Allocated	To-Date	Balance
$	$	$	$	$	$			

ESTIMATE FACTORS			COST			SCHEDULE		
Total Labor	Materials	Vendors	Budget	To-Date	Balance	Allocated	To-Date	Balance
$	$	$	$	$	$			

Implementation	Name/Title	Name/Title	Name/Title	Name/Title
BILLING RATE	$	$	$	$
% TIME	0%	0%	0%	0%
Manufacturing				
Set-up				
Equipment				
Tools				
Labor				
Marketing				
Marketing Plan				
Market Research				
Launch				
Advertising				
Sales				
Customer Support				
Distribution				
Other				
Subtotal-Time				
Subtotal-Cost	$	$	$	$
Grand Total-Time				
Grand Total-Cost	$	$	$	$

Expenses/Vendors

ESTIMATE

		Notes
Color Copies		
Color Transfers/Other Color Processes		
Computer Prints		
Courier Services/Shipping		
Fabrication		
Illustration/Other Rendering		
Legal Fees		
Local Messengers		
Modelmaking		
Photographic & A/V Supplies		
Photographic Prints		
Photography		
Portable Media (Disks, CD's, etc.)		
Presentation Materials		
Prototypes		
Reference Materials		
Reproduction		
Research & Database Services		
Research Materials		
Scanning		
Shipping		
Special Supplies/Materials		
Telecommunications		

ESTIMATE FACTORS			COST			SCHEDULE		
Total Labor	Materials	Vendors	Budget	To-Date	Balance	Allocated	To-Date	Balance
$	$	$	$	$	$			
$	$	$	$	$	$			

COST		
Budget	To-Date	Balance

Expenses/Vendors		ESTIMATE
		Notes
Travel		
Airfare		
Ground Transportation		
Lodging		
Meals		
Entertainment		
Other		
Total	**$**	

| | COST | |
Budget	To-Date	Balance
$	$	$

Credit Reference

New and unknown clients are usually a welcome challenge to a design firm. They may bring stimulating problems with opportunities for interesting design solutions. They may develop into long-term creative and financially rewarding relationships. They may also bring financial havoc. Whether a client is simply new or entirely unknown, a serious look at the newcomer's financial history would be a prudent first step toward deciding whether or not to spend the time and effort needed to produce a detailed proposal.

In the case of a new but not unknown client, a call or two to the vendors and individuals known to have business relationships with the client might produce the necessary information.

In the case of an entirely unknown prospective client, the primary way to obtain financial information is to ask the prospect for a list of credit references. Designers are often reluctant to go to this extent to protect their interests; however, they should remember that they have to furnish the same information to obtain credit with vendors and suppliers. Large or small, vendors and suppliers, rarely, if ever, open accounts without first verifying credit worthiness.

It is not enough simply to ask the client to fill out this form. It is necessary to actually contact the references and ask about the prospect's credit status. Some credit agencies may require a fee for this service. Generally, neither banks nor individuals will comment on the exact dollar worth of the prospect (which in any case is not the issue). What they are able to provide is information about the prospect's cycle of payments and general financial history. Be prompt in starting credit inquiries, since the references may take some time to respond or they might require a written request. At the same time, the client may want to move ahead with a project.

In case the prospective client is unable to furnish any credit information and the designer chooses to take the assignment, the only recourse for self-protection is to require that the client pay a substantial portion of the fee in advance, with the balance due at regular billing intervals. Reimbursable items should be billed with regular frequency and payment should be required within a specified number of days.

Filling in the Form

Fill in the date, the name of the company or individual, the billing address, telephone and fax numbers and contact name.

The client fills in the information about the company (or self), as applicable. The client also fills in the names, addresses, telephone and fax numbers, account numbers and/or contact names of the references he or she prefers to list. The client, or an authorized representative, signs and dates the form on the bottom, below the statement granting the design firm permission to run a credit check. The "notes" column is reserved for whoever in the design firm is running this process to jot down information as it is received—whether by phone or credit report. If it is a written report, positive or negative responses can be indicated generally in the "notes" column and a copy of the letter attached to this form. File the completed copy of this form.

Using the Form

The most important questions to ask when checking the prospect's credit:

❏ How long has the reference done business with the prospect?

❏ What type of business relationship has the reference had with the prospect?

❏ Has the reference extended credit to the prospect? How much? When?

❏ What is the maximum amount of credit the reference would extend to the prospect?

❏ How many days does the prospect take to pay bills?

❏ Does the prospect pay all invoices in full as presented, or does it pay on account (in small but regular payments over an extended period of time)?

Ask if the reference would have any reservations about extending credit to the prospect now and in the amount of the anticipated project billing.

Credit Reference Form

Individual or Company Name _____

Billing Address _____
_____

Date _____
Phone _____
Fax _____
Contact Name _____

Company

Years in business _____
Number of employees _____
Number of locations _____
Business type _____
   Private _____
   Incorporated _____
   Partnership _____
Federal ID Number _____

Individual

Employer Name _____
Address _____
_____
Telephone _____
Years with current employer _____
Home: own/rent _____
Years at current home address _____

Credit Agencies (Name and Address)	Telephone Number	Fax Number	Reference Number	Notes
1.				
2.				
3.				

Banks (Name and Address)	Telephone Number	Fax Number	Account Number	Contact	Notes
1.					
2.					
3.					

Trade References (Name and Address)	Telephone Number	Fax Number	Account Number	Contact	Notes
1.					
2.					
3.					

Personal References (Name and Address) **Telephone Number**   **Fax Number**   **Notes**

1. _____   _____   _____   _____
_____   _____   _____   _____

2. _____   _____   _____   _____
_____   _____   _____   _____

3. _____   _____   _____   _____
_____   _____   _____   _____

By the signature below, authorization and permission is granted to contact the references listed above for the purpose of verifying available credit information about the company and/or individual named above.

_____   _____   _____
Company Name   Authorized Signatory   Date

New Product Decision Matrix

In most instances, industrial designers are commissioned to improve, augment, and/or redesign parts of existing products. This section addresses new, possibly entirely original product development. A meticulously detailed and well-coordinated multidisciplinary evaluation of a proposed industrial design project is of paramount importance to its potential success. An understanding of the organization's strategy, its audiences—internal and external—as well as its capabilities, resources, and financial situation will reveal either an affinity or antipathy to the proposed project.

To start, an examination of the new product's fit with the business strategy of the organization should take place. Further, if the firm enjoys success as a recognized brand, it would be good to know if the new product further enhances or detracts from the established positioning of the firm's brand. It is possible that an organization may wish to move away from its traditional brand position, but in that event, it would be a good idea to first make that shift, and support it with the new product introduction. The other way around—relying on a new product to signal the important company change is a risky proposition unless preceded and followed by a strong new brand identity campaign.

It can't be overstated that this exercise should take place at the earliest stages of planning the development of a new product. Additionally, by involving all of the key people who will be involved in the new effort, it becomes possible to build the kind of support and enthusiasm necessary to sustain a long-term product development project. It's the people who take a product from concept to production and launch who are most critical to its outcome.

An understanding of industry issues—the economic climate, industry conditions, market conditions, needs, and expectations—will often reveal market opportunities that should be considered. A thorough audit of competitive products and their companies is a very useful way of finding out which product features to emulate, or improve upon and which to avoid.

Using the Form

The purpose of the new product decision matrix is to identify all of the individuals who will be involved in bringing a product to life. It's a way to make sure that the entire cast of characters is brought in early in the process—partly as a way of building support for the project, but even more importantly to ensure that they are able to make contributions based on their areas of competence at a point when early intelligence can eliminate or mitigate problems that are sure to arise at later stages.

It is also a clear and direct way of assigning responsibilities. The people selected to be members of the team will know their roles, they will know their mandates. There is a tremendous benefit for all team members to know who is responsible for what. Every job should have a contact list—not only for clients, vendors and resources, but also for an understanding of the team structure, individual roles, and responsibilities. In effect, every project needs an organization chart with contact information.

The cells in the form can be used at a minimum to check off and make sure that each topic is covered by the appropriate department(s) in the company; or more extensively, it can be used to identify the key individuals who will have a role in the project. For that purpose, a "role" key is provided in the form of DENT (Decision-maker; Expert; iNput; and core Team member). Individuals sometimes play one or more roles.

The task of selecting people to participate in this exercise will vary, but ideally it should reside with the originator of the project . . . whether a senior executive officer or an engineer in R&D. In a small organization, these roles are easy to identify; in a large organization, each department has to select its representation. While availability is often a key factor, it is more constructive to assign people based upon their skills, experience, and appetite for a new challenge.

The form is organized around a series of questions that address major topics. The questions are intended to stimulate discussion of the many different issues within each topic area.

To address corporate strategy, the question is "Why us?" The question ties back to fundamental corporate strategy, such as vision, mission, and positioning. The new endeavor is examined against its fit with the company's brand identity, and current and future customers. These are, for the most part, questions that senior executive leaders have to answer, since ultimately they are invested with the power to direct and drive the company into the future.

For market analysis, the question is "Why this product now?" Individuals in several areas can speak to this question. What are the economic, industry, and market conditions that favor the introduction of the new product? Naturally, marketing staff will be involved, but also others who are deeply knowledgeable in the areas that the product will live in—whether it's technology or tabletop design.

To gain understanding of the prospective product, we ask, "What is it and who will use it?" Most fundamentally, everyone has to know the product's features—its purpose, capabilities, and its role in consumers' lives. Then, a clear picture of the consumer must be drawn: where they're located, what their tastes and interests are, and how much they might have to spend on the product. Which, of course, points to the issues of pricing, distribution channels, and sales volume targets.

In order to make certain that all resources are in place, and to identify those that aren't, the question is, simply, "Do we have what we need to do it?" Every resource should be identified, including all of the people who will bring the product to life, the availability of financial capital, necessary technology, tools and equipment, as well as facilities.

To avoid troublesome, time-consuming, and expensive predicaments and delays, the form includes a list of regulatory considerations that must be understood early in the process to make sure that there is compliance with rules and regulations of law.

New Product Decision Matrix

Key
D Decision Maker
E Expert
N Input
T Core Team Member

CORPORATE LEADERSHIP

	Chief Executive Officer	Chief Executive Officer	Chief Creative Officer	Chief Technology Officer	R&D	Design	Engineering	Technology	Manufacturing	Quality Control
CORPORATE STRATEGY										
Why us?										
Vision										
Mission										
Positioning										
Brand identity										
Current customers										
Future customers										
MARKET ANALYSIS										
Why this product now?										
Economic conditions										
Industry conditions										
Market situation										
Market expectations										
Market need										
Market opportunity										
Competition-industry										
Competition-product										
PRODUCT ANALYSIS										
What is it & who will use it?										
Purchaser/end-user profile										
How to be used										
Features										
Technical requirements										
Durability										
Service										
Availability (distribution channels)										
Licensing										
Pricing-competitive/margins/ROI										
Value add (economies of use)										
Sales volume target										
RESOURCE REQUIREMENTS										
Do we have what we need to do it?										
Human Resources										
Leadership commitment										
Product management										
Legal guidance										
Technical knowledge										
Design skills										

	MANAGEMENT							OUTSIDE SERVICES						
	Marketing	Packaging	Sales	Legal	Finance	Human Resources	Other	Consultants	Suppliers	Vendors	Distributors	Advertising	Customer Services	Other

CORPORATE LEADERSHIP

	Chief Executive Officer	Chief Executive Officer	Chief Creative Officer	Chief Technology Officer	R&D	Design	Engineering	Technology	Manufacturing	Quality Control	
Engineering skills											
Manufacturing skills											
Marketing experience											
Sales capabilities											
Product support capabilities											
Training											
Resources-Technology, Tools & Equipment											
Model-making											
Prototyping											
Testing											
Raw materials											
Manufacturing											
Assembly											
Packaging											
Shipping											
Resources-Facilities											
Geographic location											
Space											
Layout											
Infrastructure											
Warehouse/storage											
REGULATORY CONSIDERATIONS											
International, Federal, State, Local											
Materials											
Ingredients											
Parts											
Labeling											
Environmental											
Dimensions											
Packaging											
Pricing											
Selling practices											
Consumer rights											

	MANAGEMENT							OUTSIDE SERVICES						
	Marketing	Packaging	Sales	Legal	Finance	Human Resources	Other	Consultants	Suppliers	Vendors	Distributors	Advertising	Customer Services	Other

Proposal

Reputation, recommendations, astutely targeted marketing, and self-promotion all play their roles in capturing a client's attention. A critical, final, deciding factor in the client's selection process is a clear, concise, and accurate proposal outlining the scope and schedule of work, cost of materials, fees, and terms.

In the following explanation of this form, we refer to "client" as the purchaser of services. It might be that you are a designer-entrepreneur and do not need to create a proposal. In which case the only purpose this form might have would be to help you organize an overall picture of your project. In effect, it could serve as the business plan for the project.

Clients often use proposals to compare estimates by several designers being considered for an assignment. The ability to turn around a professional document in a timely manner is useful in winning projects. Consider each proposal as your final chance to persuade the client to hire you. It is concrete evidence of your creative and organizational skills. While it is not a promotional brochure, it is representative of your aesthetic sensibility. Therefore, use your professional letterhead or a special proposal document format. Make sure it reads easily to a nontechnical person, and that it is clearly organized and impeccable in its appearance. Most important of all, respond directly to the client's request. Address all of the issues the client has raised up to this point, proving that you understand the situation and goals and can respond with ideas and recommendations that are creative and appropriate.

If information provided by the client is vague or incomplete, you are in a situation where you are expected to provide a "ballpark" proposal that gives the client an idea of approximately how much a project might cost. Such a proposal winnows out the designers to be considered seriously, at which point a more extensive proposal is usually requested. While ballpark proposals may be limited in detail, they should include a description of everything relevant that is known about the assignment, approximate fee amounts, schedule, and expense policy. Do include proposed start and end dates, show only generalized timelines, so that when more information is provided, you can clearly show how scheduling is impacted by the new specifications. Details can be generalized, but clearly state that this is a preliminary proposal and that, if accepted, additional information reflecting updated specifications will be forthcoming. It is especially necessary to include this statement because additional information and changes will have an impact on the project's final cost and schedule, and your fee should reflect that.

The most important information to include in a proposal are 1) the scope of work, i.e., an understanding of the assignment by means of a complete description; 2) your methodology, or an outline of your working process, including worksteps and deliverables; 3) fees and expense policies (billables, markups, schedule of billings, etc.); 4) a time schedule; and possibly, 5) basic terms (changes, termination, payment schedule, releases, intellectual property rights, etc.). It also helps to specify the number of concepts that will be presented and in what format; the number of rounds of revisions; and approximately how many meetings and presentations can be expected.

Most clients will want to have a more extensive contract agreement because the nature of industrial design is that there is need for a great deal of detailed information about topics such as intellectual and real property ownership, indemnification (an especially critical subject for industrial designers), and confidentiality. Additionally, when the designer is a part owner of the copyright on the product, royalty terms have to be specified. In this case, leave terms and conditions out of your proposal, since all those details will be negotiated and itemized in a master contract once the scope and fee are agreed upon.

Proposals can be standardized, such as the example shown here, or they may be written in composition style in a letter. Generally, designers use their own standard formats with a personal cover letter, which is much like an "executive summary." This summary highlights the salient points of the scope of work, approximate overall fee or fee range, and proposed completion date, along with a personal expression of the designer's interest in the project and desire to work with the prospective client.

When working with a new client, particularly on a project that is extensive or not well defined, it is best to draw up a standard contract such as the one in this book, or one issued by the American Society of Industrial Designers.

Filling in the Form
Provide all of the client information indicated, the date of the proposal, who is writing it ("by"), and give the project an easily identifiable name. You may already have all this information if you have opened a job index and job sheet for this project. In which case, simply copy and paste in the information. In the "fee information" section, fill in the total fee and amounts for out-of-pocket expenses (OOPs) and travel, if applicable. Under "scope of work," summarize the work to be done including enough information to indicate the level of complexity of the proposed project. In the "work plan" area, under "assumption" indicate whether time is based upon hours, days or weeks. Fill in the projected start and end dates for each work step and the associated cost for each phase of the proposed project. Design services are often billed in advance on a phase-by-phase basis. In case the project is canceled, it will be easier to collect the appropriate fee for completed work phases. Total the "budget" column and show either overall project start and finish dates or the total amount of work time the project will require, regardless of start and finish dates.

On the "terms and conditions" page, fill in or delete all of the blanks.

Using the Form
In the case of a well defined, small-scale project, the proposal can be written so that, if you are selected to do the job and the proposal needs no adjustments (assuming both the client and the designer are in complete agreement about the extent of the work to be done and its execution), it can be signed by both parties and become a letter of agreement. To do this, make sure that there is a statement preceding the signature:

"Your signature below authorizes (designer) to commence work on (name of project) for the fee of ($_____); and affirms agreement with the Scope of Work, Budget, Schedule, and Terms and Conditions, specified herein. Kindly return a signed copy of this Proposal to (name of designer). Upon signature of both (designer) and (client), this document shall become the Agreement between these two parties."

Leave room for your name both in handwriting and printed form and the date of your signature; and the same for the client. Send a copy with your signature and as soon as the client returns it signed and dated, you can proceed to work. Faxed signatures are binding. Additionally, the letter of agreement must contain both the client's and the designer's names in print and signature form, addresses, and the signing date, as well as specific information about the number of preliminary sketches, billing for extras and changes, payment for materials, payment of sales tax, liability, and a termination fee.

In a competitive situation price is not always the sole determining factor in winning a project. It is better to be within the range of given price quotes than to be dramatically lower or higher. An extreme departure from the general range of price quotes signals a lack of understanding of the assignment, desperation, or indifference. Do some research about fees other industrial designers are charging for similar a scope of work.

Proposal

Date _____

By _____

CLIENT INFORMATION

Company Name _____

Address (Meetings) _____

Phone _____

Address (Billing) _____

Phone _____

Cell _____

Fax _____

Primary Contact _____

Other Contact Information _____

PROJECT INFORMATION

Name _____

FEE INFORMATION				
	Fee	OOPs*	Travel**	Total
				$
*Expenses-Billable (with markup)	(list)			
**Expenses-Billable (without markup)	(list)			

SCOPE OF WORK

(Description)

WORK PLAN

Assumption: Time and rate calculations are per   [] Hour   [] Day   [] Week

	Start Date	End Date	Duration	Budget
Strategic Analysis and Planning				
Project Commencement/Pre-Planning	_____	_____	_____	_____
Intake Meetings	_____	_____	_____	_____
Audits	_____	_____	_____	_____
Client	_____	_____	_____	_____
Industry-Competitive	_____	_____	_____	_____
Market	_____	_____	_____	_____
Product	_____	_____	_____	_____
Targeted Strategy Development	_____	_____	_____	_____
Product Brief	_____	_____	_____	_____
Design Brief	_____	_____	_____	_____
Meetings	_____	_____	_____	_____
Revisions	_____	_____	_____	_____
Subtotal				$
Creative Development				
Ideation-Brainstorming/Concept	_____	_____	_____	_____
Renderings	_____	_____	_____	_____
Modelmaking	_____	_____	_____	_____
Evaluation	_____	_____	_____	_____
Strategic	_____	_____	_____	_____
Functional	_____	_____	_____	_____

Refinements/Revisions				
Prototyping				
Testing				
Client Meetings/Worksessions				
Refinements/Revisions				
Concepts Presentation				
Refinements/Revisions				
Client Sign-Off				
Design Development				
Specifications				
Testing				
Client Meetings/Worksessions				
Refinements/Revisions				
Design Development Presentation				
Refinements/Revisions				
Client Sign-Off				
Pilot Production				
Sub-Assembly				
Testing				
Refinements/Revisions				
Final Specs and Blueprints				
Contract Documents and Bids				
Bid Drawings and Specs				
Bid Review				
Client Review of Bids				
Refinement/Revisions				
Client Sign-Off				
Vendor Selection & Contract Documents				
Subtotal				$
Implementation				
Manufacturing				
Set-up				
Equipment				
Tools				
Labor				
Marketing				
Marketing Plan				
Market Research				
Launch				
Advertising				
Sales				
Distribution				
Customer Support				
Other				
Subtotal				$
Total				$

All information in this proposal is subject to the Terms and Conditions listed herein.

TERMS AND CONDITIONS	
Change Orders	Work change orders will be issued for additional scope of work and changes requested after approval of plans or commencement of work. WCOs include a description of the change/addition requested, estimated additional costs and changes to work schedules/project completion. Client's signature is required on WCOs to proceed with changes/additions.
Billable Items	In addition to the fees and costs estimated herein, costs incurred for all out-of-pocket expenses (OOPs) are billable (at cost/with a markup of x%). Wherever applicable, state and local sales taxes will be included in Billable Items.
Purchasing	All purchases made on client's behalf will be billed to client. In all cases, such prices will reflect a (markup) of ___%. Charges for sales tax, insurance, storage, shipping & handling are additional to the price of each purchase. In the event client purchases materials, services, or any items other than those specified by the designer, the designer is not liable for the cost, quality, workmanship, or condition of such items.
Deposits	A 50% deposit is required on all orders made on behalf of the client, prior to the placement of such orders. The balance is due upon delivery. Client is liable for all costs and expenses for any items canceled by the client, but that have already been ordered by designer, unless such items can be cancelled at no charge.
Custom Orders	Client is fully responsible for paying all costs for special orders canceled by client once manufacturing has started. The cost of special orders that are canceled by client following approval to purchase, is fully dependent upon the terms and conditions put forth by the manufacturer, supplier, or vendor of the item(s).
Warranty	Designer makes no additional warranty, guarantee, or any other assurances, other than those provided by manufacturers, suppliers, or vendors of products and/or services.
Other Contracts	Designer is not responsible for any contracts that the client enters into directly with contractors, suppliers, manufacturers, or vendors for goods and/or services, whether or not such sources are recommended by the designer.
Purchase Orders	Designer is responsible for those purchase orders issued directly by designer. Client is fully responsible for all purchase orders issued by client. Responsibility includes, but is not limited to, errors, omissions, pricing and scheduling.
Schedule of Payment	Hourly rate: Regular billing periods (bimonthly, monthly) based on hours consumed or periodic approval points. Flat fee rate: ___% upon signing of project agreement; ___% upon concept development sign-off; ___% upon design development sign-off; ___% upon product launch. Invoices are payble upon receipt.
Intellectual Rights and Ownership	Upon full payment of all project fees and costs due, the following rights to the use of the designs, including drawings, specifications, models, and prototypes, transfer to client, as noted:

Indemnification	Designer makes no additional warranty, guarantee, or any other assurances, other than those provided by suppliers, vendors, or manufacturers of products and/or services.
Termination	Client and designer may terminate project based upon mutually agreeable terms to be determined in writing either prior to signing of this proposal, or within the final client-designer contract.
Term of Proposal	The information contained in this proposal is valid for 30 days. Proposals approved and signed by the client are binding upon the designer and client beginning on the date of agreement.

Your signature below authorizes (designer) to commence work on (name of project) for the fee of ($_____); and affirms agreement with the Scope of Work, Budget, Schedule, and Terms and Conditions, specified herein. Kindly return a signed copy of this Proposal to (name of designer). Upon signature of both (designer) and (client), this document shall become the Agreement between these two parties.

Designer Signature _____   Print Name _____   Date _____

Client Signature _____   Print Name _____   Date _____

Billable Out-of-Pocket Expenses	**Budget**
Color Copies	
Color Transfers/Other Color Processes	
Computer Prints	
Courier Services/Shipping	
Fabrication	
Illustration/Other Rendering	
Legal Fees	
Local Messengers	
Modelmaking	
Photographic & A/V Supplies	
Photographic Prints	
Photography	
Portable Media (Disks, CDs, etc.)	
Presentation Materials	
Prototypes	
Reference Materials	
Reproduction	
Research & Database Services	
Research Materials	
Scanning	
Shipping	
Special Supplies/Materials	
Telecommunications	
Travel	
Airfare	
Ground Transportation	
Lodging	
Meals	
Entertainment	
Other	
Total	$

Time Sheet

Regardless of whether a project is billed on a flat fee, retainer, or an hourly rate basis, it is essential to know exactly how much time all staff members spend on every project. Additionally, it is useful to know how non-billable time is being used. Naturally, it is important to have an accurate record for jobs billed by the hour, mostly to ensure that all billable time is reimbursed, and partly in the event the client requests an audit of time-keeping records.

Time records should also be kept for projects that are billed on a non-hourly basis. A flat fee, whether based on a percentage of the cost of the project, royalties, or a highly defined estimate of costs and time, has to cover at least the amount of time to be spent on a project, with a markup to include profit and overhead expenses. (Non-billable costs such as rent, utilities, insurance, and other overhead items are discussed under the job sheet section.) To approximate the time to be spent on a project, the industrial designer needs to think the assignment through all of its stages, its scope of work and complexity and what, if any, kinds of outside services will be required, such as software, electronic, or mechanical development, among others. Also, it is good to take into account the nature of the client—are they known to conduct numerous and lengthy meetings, do they require many sketches, do they sign-off on designs in a timely manner, do they tend to request changes in mid-stream? A high-maintenance client can consume valuable time and human resources.

Time is the variable in determining fees and the most critical key to profitability on a flat-fee-based assignment. The more exact the understanding is of the technical aspects and client contact required, the more accurate the estimation of the time required. As a safety device, some industrial designers customarily add 15 percent to whatever number of hours they estimate will be needed to complete the assignment—this is also called a contingency factor.

Filling in the Form

Each staff member fills in his or her name, the current month, dates covered by the time sheet (usually covering a week's activities) and the year. Fill in the "project number" column so that whoever is posting time sheets to job sheets does not have to refer to the jobs index every time a job name appears. Mid- to large-scale industrial design firms now set up these tracking documents on their Intranets and can automate the transfer of data from timesheets to job sheets. You could also set up job sheets for non-project types of activities, to capture time spent on general administration, maintenance, sick days, etc.

Fill in the "project name" as listed on the jobs index. Indicate the day of the week or date under "date." Specify activities in code under the headings of "phase code," "activity code," and "detail code." Indicate the number of hours worked on each job every day. It is important to note the "total billable" time, because project agreements may vary in regard to which activities are billable, and which are not. The "notes" area is useful for additional comments.

The "notes" column can be used to indicate the percentage of billable versus non-billable hours spent (also called utilization). Or, it can be used to show other calculations, such as the number of overtime hours, or marketing versus project time, or other kinds of staff time information that the design firm would like to track.

The staff or project supervisor should sign or date and sign the "approved" line.

Using the Form

The coding system shown here is a suggested format that follows the general outline of the business system suggested throughout the business forms part of this book, including the estimate and job sheet forms; you may decide to either simplify, show greater detail, or use other coding symbols of your choice.

The advantage of this kind of breakout of phases, activities and details in separate columns, is that in Excel, you can sort information by columns and analyze staff members' time spent on various activities.

Above specific grades of employment, especially positions that are considered "professional," industrial design firm employees do not get paid overtime. The reason you should track it anyway is that on hourly jobs, while you are billing the total time worked (regardless of whether it is regular or overtime), you are incurring additional costs for staff that is working overtime. Such costs might include overhead items such as utilities (heat and air conditioning) and variable costs such as food and transportation for late-night work sessions. Knowing how much overtime is incurred is also helpful in determining if a staff person

may be entitled to some additional time off, or "comp" time, particularly if the overtime is incurred to meet client demands rather than for slow work habits.

All are required to fill out a time sheet—and for the sake of billing, principals and other senior personnel need to be accountable—should fill out their time sheet daily. It is easier to remember exactly how much time is spent on specific activities at the end of the day than at the end of the week.

Time sheets should be checked and signed by project leads or senior designers, as appropriate to the management style of your firm. Principals and senior staff usually turn over their time sheets directly to the person who does the posting.

Time sheets should be posted weekly, so that no more than a few days are needed to prepare billing when an assignment is complete or ready for the next phase of billing. Also, some clients like to get weekly time accounts of their project's activities.

Time Sheet

Name _____   Dates _____

Month _____   Year _____

Billable Codes

Phase Codes	Phases	Activity Codes	Activities	Detail Codes	Detail
1	Pre-Plan	A	Audits	00	Not a Detail Item
2	Strategic Analysis	B	Budget/Schedule/ Proposal	01	Presentation Preparation
3	Concept Development	C	Client Relations (Meetings/Presentations)	02	Travel Time
4	Design Development	D	Design Drawings	03	Airfare
5	Pilot Production	M	Model-making	04	Ground Transportation
6	Manufacturing	P	Prototype Development	05	Meals
7	Marketing	Q	Quality Control	06	Lodging
8	Sourcing	R	Refinement	07	Product Brief
9	Production	S	Sourcing	08	Design Brief
10	Launch	T	Testing	09	Materials
11	Distribution	V	Travel	10	Parts
12	Product Support	W	Copy Writing/Editing/Proofing	11	Color Studies

Nonbillable Codes

Use these codes in the "Phase" column

GA	General Administration
MN	Maintenance
ED	Education/Training
PM	General Promotion/Marketing
VA	Vacation
HD	Holidays
SK	Sick Days
EA	Excused Absences (July Duty, Family Leave, Etc.)

CODES

Job Number	Job Name	Phase	Activity	Detail	Mon	Tue	Wed	Thu	Fri	Sat	Sun	Total	Total Billable	Notes
Totals														

Signature _____   Approved _____

Date _____   Date _____

Project Status Report

Designers use project status reports for long-term and complicated projects, where they are invaluable in keeping clients informed. A status report includes information about what is presently going on, key dates, who is waiting for what, from whom, and when. Used regularly, these reports help maintain communications between designer and client and can also be helpful in keeping subcontractors tuned into the rhythm and direction of the project. For short-term, less complex assignments, project status reports can be used to prod a client out of a stalled situation. This report can also be particularly useful to non-visual or inexperienced clients because it gives them a clear picture of what is expected to happen and in what order. Using standard professional terminology also allows the client to become familiar with the language, as well as the process of industrial design. Finally, a status report helps to identify those individuals who have responsibility to carry out specific tasks. It is a clear, but non-confrontational way of getting people to respond to specific issues and needs.

Filling in the Form

Fill in the job name, number, and date of the report. In the "to" and "copies" fields, write in the name of the primary client contact who will be receiving this report, or the business name of the client if it is to be circulated widely, as well as the list the names of additional people who should be receiving copies. (This format may also be used to summarize the minutes of meetings, in which case the date of the meeting should be indicated.)

There are two ways to use this form: 1) For short specific tasks that correspond to the list of phases, check mark the applicable phases, and fill in the items and other information in the spaces provided; or 2) if the need is to discuss one or two topics in greater length and detail, keep only the phase appropriate to the discussion, delete all the others, and use the entire space available on the sheet to outline the points to be made.

Using the Form

Someone on the design team should always take notes during client meetings. The ability to articulate the client's needs and ideas briefly, along with a written record of critical decisions, serves as a great advantage in conceptualizing and moving a project along.

When it is useful to keep subcontractors (such as software developers, model-makers, packaging designers) apprised of project developments, this form can be addressed either directly to them, or they can be included in the general distribution of copies. Very often, subcontractors are either directly affected by specific changes on a project or they simply need to know whether the work is proceeding according to schedule.

Industrial design projects have critical deadlines and it is important for all involved to know how their contribution fits into the overall scheme. It cannot be overestimated that clear, timely, and accurate communications are essential to the success of team-based projects.

Project Status Report

Job No. _____   To _____   Date _____

Job Name _____   Copies _____   From _____

Phase	Item	Status	Action Required	From	To	Date Due
Strategic Analysis and Pre-Planning						
Creative Development Concept Design Pilot Production Testing						
Implementation Manufacturing Plan Production Development Trial Production Quality Control						
Marketing Advertising Sales						
Other						

Work Change Order

In the course of virtually every industrial design project, it is inevitable that either clients will request changes, or that in the development process discoveries are made that require a change in the scope of work. Whether such changes occur early in the life cycle of the project or later, it is very important to document those requests that affect either the project budget and/or schedule for completion. In particular, any changes to items that were previously approved should be noted: first, to verify the exact nature of the change; and second, to justify any additional (or reduced) billing and scheduling that the change necessitates.

A brief and simple form is easier for clients to read, sign, and return than addenda to contracts, or even letters outlining the new client instructions. It is extremely important to follow through on obtaining signed copies of this form.

Filling in the Form

Fill in the name of the client and the project. Include the name of the person who has initiated the change request. Also, include the project number, the date, and a number for the change order. The work change order number can be linked to the individual project number, or there could be an office-wide work change order system, like purchase orders.

Check the stage during which the change is being requested. In the "work change description" area, describe what is being changed. Under "cost change" indicate the additional or reduced amount. In general, negative numbers in bookkeeping are indicated by ( ) marks. Also, fill in the change to the schedule—expanded or collapsed. If a lengthy description is required, attach additional pages to this form. Make sure to also indicate if drawings are attached.

Using the Form

Keep close track of all additional time and costs incurred by changes in project scope or specifications. If the client questions additional changes, these work change orders serve as proof that additional work took place either at the client's behest or with the client's approval for changes initiated by the designer.

Keep this form in the job file. Jot down the form's date, number, and amount on the job sheet (Form 2).

Work Change Order

Client _____

Change Order Number _____

Project _____

Date _____

Work Change Requested By _____

Job Number _____

Phase

Pre-Planning _____

Concept Development _____
_____

Design Development _____

Implementation _____

Other _____
_____
_____
_____

Work Change Description	**Cost Change**	**Schedule Change**

This is not an invoice. Revised specifications to work-in-progress represents information that is either different from that which the original project budget and schedule were based upon, or follows after client's approval to the stage of work in which this (these) item(s) appear(s). Changes in time and cost quoted here may be approximate, unless otherwise noted.

Your signature below will constitute authorization to proceed with the change(s) noted above.
Kindly return a signed and dated copy of this form to:
_____

The information contained in this Work Change Order is assumed to be correct and acceptable to client unless designer is other wise notified in writing within _____ days of the date of this document.

Authorized Signature _____

Print Name _____

Date _____

Client Approval

Serious confusion and error can be avoided by instituting a process that gives the client the opportunity to sign off on both general and specific project elements. Again, "client" refers to the person who is charged with ensuring the correct outcome of the design process and has authority to sign financial agreements, such as contracts and large purchase orders—it might be a corporate or an entrepreneur client, it might be an executive or division head within your company. Additionally, if changes are made following client approvals, the designer is in a significantly better position to issue work change orders (Form 10) and to defend changes in billing and scheduling. It is also an opportunity to catch issues that need additional attention and possible redesign. It is always better to have a client that is in sync with the project details and can offer clear guidelines at appropriate times. Although it is noted on the bottom of this form, it is important to stipulate in the project contract that changes following client approvals and sign-offs are subject to changes in budget and schedule.

Filling in the Form

Fill in the client and project names, and the project number. If you are using the form in its entirety and will keep it as a running approval list, instruct the client (or highlight the appropriate rows) as to which items you are requesting to have approved. Show the client where to sign and where to print their name and the date. If you prefer to have separate forms for different phases or items, delete the unnecessary information and proceed to

have it signed and dated. If you are mailing the form, make sure to keep a copy indicating when you sent it out. Attach a note to the form specifying when you need to have sign-off or approval in order to be able to proceed in time to stay on schedule. Be diligent in collecting these sign-offs. They will save you major headaches when a client claims that you haven't delivered what was expected.

Using the Form

While this form has a moderate level of detail, it is easy to reformat by adding or deleting rows that are either necessary or irrelevant to the project or client approval. It could be enough to maintain the "approved to proceed to next phase" lines in each phase, along with the final project sign-off.

The trick, of course, is to get clients to sign off. Some client representatives are loath to sign their names to anything. The best approach to this problem is to state in the project contract that written client approvals to proceed from phase to phase is a condition of bringing the project in on schedule. It would also help to note on the project status report (Form 9) that client sign-off is pending.

If the client or the client contact is unwilling to physically sign the approval form, it might be necessary to issue a note to the effect that, "In the absence of written client approval, the information contained in the attached client approval form is assumed to be correct and acceptable to the client unless the designer is notified in writing within ____ days of the date of this notice."

Client Approval

CLIENT
Name _____
PROJECT
Name _____
Job Number _____

Phase	Item	Print Name	Signature	Date
Strategic Analysis and Pre-Planning	Product Brief Design Brief Approved to proceed to next phase			
Ideation/Concept Development	Concept Drawings Models Strategic Evaluation Functional Evaluation Concept Sign-Off Approved to proceed to next phase			
Design Development	Design Drawings Prototype Strategic Evaluation Functional Evaluation Testing Specifications Design Development Sign-Off Approved to proceed to next phase			
Pilot Production	Sub-Assembly Testing Final Specs and Blueprints Contract Documents and Bids Vendor Selection & Contract Documents Approved to proceed to next phase Final approval			

Phase	Item	Print Name	Signature	Date
Implementation	Manufacturing Plan-Approved Marketing Plan-Approved Market Research-Approved Launch Plan-Approved Advertising Plan-Approved Sales Plan-Approved Distribution Plan-Approved Customer Support Plan-Approved			

NOTE: Changes made following client approvals to proceed and sign-offs on specific elements are subject to changes in budget and schedules. In such cases, a Work Change Order will be issued that will specify any additions or reductions to budgets and schedules.

Transmittal

The transmittal form is most frequently used as a kind of cover identification for enclosures, attachments and any other kind of material being disseminated within or outside of the firm. The advantage of having a multiuse transmittal form is that it eliminates the need to create individually written cover notes every time material needs to be circulated. Also, the information on the form is comprehensive, thereby uniformly communicating the necessary facts about the accompanying items. These forms may be printed in hard copy on the designer's letterhead, or kept as a digital document that is generated as needed. Make sure your firm's name, address, telephone and fax numbers are easily readable. Run a test copy through your fax machine to make sure that the copies it generates are sharp and clear. Avoid the temptation to use tiny type and soft colors on this form.

Filling in the Form
Fill in the name of the recipient, his or her company's name, and telephone and fax numbers. Include the name of the person sending the transmittal ("from"), the date, and the project number, if applicable. List the names of additional people who will also be receiving a copy of this form. Check the reason "for" sending the accompanying material. In the "via" section, check the means by which this communication is being sent. For "enclosed" checkmark the type of attachment or enclosure. In the "disposition" area indicate whether the material is to be returned, kept or distributed. Use the space left for "remarks" for additional messages and instructions.

Using the Form
Keep copies of transmittals having to do with active project work. It may be helpful in identifying the location of missing items or paperwork.

If material of significant value is being transmitted (such as a prototype), obtain adequate insurance to cover loss and damage in shipment.

If the enclosed material being transmitted is valuable, indicate who is responsible for its loss or damage—including the time during shipment.

Transmittal Form

To _____   From _____

Company _____   Date _____

Phone _____   Job Number _____

Fax _____

Copies to: _____

For: _____

❏ Review   ❏ Files   ❏ Information

❏ Approval   ❏ Distribution   ❏ As Requested

Via: _____

❏ Fax (Number of pages, including transmittal/cover)

❏ Messenger   ❏ Pouch   ❏ Interoffice

❏ Courier Service _____

❏ Freight Forwarder _____

❏ US Mail (regular)   ❏ US Mail (express)

Enclosed: _____

❏ Text Document   ❏ Model   ❏ Presentation Material

❏ Drawings   ❏ Prototype   ❏ Promotion Package

❏ Other: ❏ Part/Material/Sample   ❏ Publication

Media: _____

❏ Hard Copy

❏ Digital   Media Type   Disk/Folder/File Name(s)

                        ❏ CD-ROM   _____

                        ❏ DVD   _____

                        ❏ Jaz Disk (Mac) (PC)   _____

                        ❏ Floppy Disk (Mac) (PC)   _____

                        ❏ Video (VCR) (Beta)   _____

                        ❏ Other   _____

Disposition: _____

❏ Kindly Reply   ❏ Return   ❏ Keep   ❏ Distribute

Remarks: _____

_____

_____

Traffic Log

It is a basic need of every design office to know the whereabouts of materials that have left the premises. In a relatively small industrial design firm, it may be enough to have one overall traffic log to track the comings and goings of documents, drawings, models, samples and so on. The following form is designed to meet this kind of need. With very minor adjustment, it can also be adapted for use on individual large, complex and long-term projects. To do this, simply include the following information in the space under the form's title, as follows in this example:

Traffic Log

CLIENT

Name _____

PROJECT

Name _____

Location _____

Job Number _____

Then, continue with the rest of the headings, as provided in this form.

Filling in the Form

Fill in the date the item is sent, to whom (include both the name of the company and the street address shown on the shipping label or envelope) and very briefly what the item is. Note the media (as indicated on the transmittal form) and what (if any) other enclosures are included. Indicate the messenger, courier, or shipper under "via," and if the item is supposed to be returned, place a checkmark under "return." If applicable, fill in the project number and who requested that the item be sent, if that person is not managing the log.

Using the Form

Transmittals do not take the place of this log. Digging through copies of transmittals will not provide the information you need quickly to locate where an item was sent, when, and by whom.

This form should be located in or near the design office's traffic area and can be filled in by individual staff members; or, it can be the responsibility of one person such as a receptionist—in which case all staff members need to transmit the necessary information to this person. In large firms, individual departments maintain their own traffic logs.

Traffic Log

Date	Time	Job Number	Sent By	Sent To

Description (Item)	Via	Due Back	Returned	Received By

Expense Report

Expense reports are used to quantify expenses incurred by employees. It's an easy way to organize piles of receipts and assures that billable expenses are charged back to clients. Some clients accept the report without copies of the original receipts, which makes billing easier for the designer. Other clients insist on a spreadsheet that summarizes the expenses by categories, while others require original receipts, in which case it's important to keep copies on file.

Filling in the Form

Fill in the identifying information at the top of the form. Expense reports should tie back to specific projects or company activities (such as marketing, education, special events, etc.) Tape the original receipts on 8.5″ × 11″ sheets of paper and number the receipts. Reference each receipt by its number to the form. Fill in the location and date of the expense, what it is, place the total amount in the billable or non-billable columns; put in the exchange rate if the expenses occurred as a result of foreign travel; and place the amount again in the correct category column. Original receipts for out-of-pocket purchases and/or travel must accompany the expense report.

Using the Form

Every organization has rules concerning reimbursement for employee-generated expenses. It is best to codify these rules and make them available to all employees who incur expenses on behalf of the company, so they have a clear understanding of these guidelines.

Expense Report

Expense Report
  EMPLOYEE
Name _____
Department _____
  CLIENT
Name _____
  PROJECT
Name _____
Job Number _____

Location	Receipt Number	Receipt Date	Item	Billable	Non-Billable	Exchange Rate	Ground Transportation
				$	$	$	
				$	$	$	
				$	$	$	
				$	$	$	
				$	$	$	
				$	$	$	
				$	$	$	
				$	$	$	
				$	$	$	
				$	$	$	
				$	$	$	
				$	$	$	
				$	$	$	
				$	$	$	
				$	$	$	
				$	$	$	
				$	$	$	
				$	$	$	
				$	$	$	
				$	$	$	
				$	$	$	
				$	$	$	
				$	$	$	
				$	$	$	
				$	$	$	
				$	$	$	
				$	$	$	
				$	$	$	
				$	$	$	
				$	$	$	
				$	$	$	
				$	$	$	
				$	$	$	
				$	$	$	
Total Amount				$	$	$	
Total Mileage							

EMPLOYEE
Signature _____   Date _____

APPROVAL
Signature _____

Print Name _____

	Airfare	Mileage	Meals	Lodging	Entertainment	Nonbillable	Supplies		Other	Total
	$	$	$	$	$	$	$	$	$	$

Estimate Request

FORM 15

Whenever you have to select sources and services (parts, materials, special technology development, packaging, etc.), it becomes necessary to request estimates from eligible and appropriate suppliers. To avoid confusion in evaluating these estimates, it is useful to provide the competing suppliers with exactly the same description of the work to be performed. This form may also be used to back up a verbal quote when the supplier has been selected without preliminary screening. Although it is clearly a quotation of approximate fees and costs, the estimate request helps to control significant variations in vendor bids. As noted elsewhere in this book, the difference between an estimate and a bid is that an estimate is an approximate calculation and a bid is a specific proposal whose numbers are binding, unless the specifications (specs) that led to the bid change.

Filling in the Form

Fill in the name of the client and project, the date, the job number, the name of the person requesting the estimate and his or her contact information. Fill in the name and address of the vendor/supplier. Fill in "specifications/description" as completely as possible; include as much information as necessary to accurately estimate the work to be done, or items to be purchased.

Leave blank the spaces for estimate, subtotal, shipping/handling, tax, total and deposit required. These spaces are for the vendor/supplier. Fill in the delivery date. The vendor/supplier should sign the estimate, provide contact information and include the date the form was signed.

Using the Form

When requesting an estimate, fill in the heading and the specifications/description part of the form; make as many copies as needed to distribute to the prospective bidders (plus one for your files, of course); and customize the vendor/supplier information individually as appropriate. This way it is certain that everyone has exactly the same instructions on which to base their estimates.

If necessary, send a copy of all the completed estimate requests to the client. Include information that will have an effect on the estimate, even though it is not included in the technical spec. For example, if there are complicated shipping requirements, or if the assignment will require the vendor to travel, or any other significant service or condition that is in addition to the basic work requested.

Keep this form in the job file. Jot down the estimate from the selected supplier in the "summary of costs" area on the project's job sheet.

Estimate Request Form

Client _____

Project Name _____

Job Number _____

Date _____

Request By _____

Phone _____

Fax _____

E-Mail _____

Supplier Company

Contact Name _____

Address _____

Phone _____

Fax _____

E-Mail _____

Specification/Description

Item Number	Specifications/Description	Delivery Date	Est. Quote	Total
1				
2				
3				
4				
5				
6				
7				
8				
9				
			Subtotal	$0.00
			Shipping/Handling	
			Tax	
			Total Estimate	$0.00

Deposit
Required _____

Date _____

Notes _____

_____

_____

_____

_____

Supplier Signature _____   Date _____

Print Name _____

It is understood that while the estimated costs are approximate, final billing will be adjusted according to specific instructions provided in a purchase order or contract. Kindly fill in the information requested, sign, date, and return a copy of this form by _____ Thank you.

Estimate Log

The estimate log is the place to list the results of the estimate requests made for the manufacture, fabrication or purchase of special items for which there are no standard prices that can be found in catalogs and other product information sources. The estimate request is recommended here to use in gathering multiple estimates.

Filling in the Form

Fill in the client name, the project information, including the project's name and the job number. List the date the estimate is being logged (it is a good idea to log these as soon as they come in), the name of the supplier (or contractor, vendor, etc.), and the date the original estimate request was sent. Include the item number from the estimate request form, the item's name and the quantity requested. From the supplier's response, list the unit price quoted, as well as the subtotal of all the items and any other costs such as shipping and handling and taxes, if applicable. Make sure to note if a deposit is required (and how much), as well as the delivery date quoted. In the last column, indicate which supplier was chosen for the assignment.

Using the Form

When there are several contractors and vendors who can provide similar products and services, and the designer does not have a favorite to work with, it is time to solicit estimates. Or, if the designer is unfamiliar with the cost of a special item, it is very helpful to learn the possible ranges in price, quality, and delivery schedules of such items.

Certain clients, most often institutional, corporate and governmental, will require a minimum number of bids for the manufacture and purchase of all project items. It is very important to be familiar with the guidelines provided by these clients as they are usually quite stringent in maintaining these standards and usually reserve the right to audit records.

In the event similar items or services are required later for other projects, and a known contractor or vendor is no longer available, here is a quick way to look up possible alternatives or replacements.

Estimate Log

CLIENT
Name _____
PROJECT
Name _____
Job Number _____

Log Date	Supplier Name	Est. Request Date	Item No.	Item	Quantity	Unit Price	SubTotal	Shipping/ Handling	Tax	Total Estimate	Deposit	Delivery Date	Selected Vendor

Contractor Log

In many instances, industrial designers work with their own team of crafts- and tradespeople. Often clients will select a designer who is known to have access to teams of workers who are excellent and/or unique in their trades. In this sense, industrial designers act not only as creative directors, but also as general contractors. They are responsible for the selection of contractors, establishing secure contracts with them, and for supervising their work. The purpose of the contractor log is to have one place to look for the list of all of the contracts generated by a project.

This log is meant to be used only for items requiring technology, software, electronic and/or mechanical, and other specialized development. This is not intended to be a list of purchases.

Filling in the Form

Fill in the client name, project name, and the job number. Indicate the trade or specific element for which the contractor is engaged under "item." List the contractor's company name, and the name of the primary contact, including his or her phone/fax numbers. In the "bid documents" area, indicate what specs and drawings were provided, and the date and number (if applicable) of the contract. Under "pricing," list the purchase order number and its total amount, if one was issued. List any work change order numbers and the final agreed price under "invoice total." For "schedule," list the date the work was ordered (this could be the date of the purchase order), the delivery or due date, and when the work was actually received or completed.

Using the Form

While the estimate log (Form 16) is a list of all of the bids received for the project, the contractor log contains only the information about the selected contractors.

The contractor log is also a good reference for future projects that require similar items and services. It is also helpful to be able to look up pricing and scheduling for special custom items.

Contractor Log

CLIENT

Name _____

PROJECT

Name _____

Job Number _____

Item	Contractor	Contact	Bid Documents		P.O. #
			Specs	Drawings	

	Pricing			Schedule			
	P.O. Total	Change Orders	Invoice Total	Start	Inspection	Delivery/Due	Received/Completed

Purchase Order

The purchase order serves as a written notice to vendors, manufacturers, and other suppliers, including freelancers and independent contractors, to begin work on a specific assignment or to deliver goods. Most vendors will not proceed without a written purchase order. For the design studio, the purchase order form is handy in two ways. First, it is a record of when goods are ordered, from whom, and when delivery can be expected. Second, when invoices are being checked, purchase orders are useful for verifying precisely what was ordered, and at what price. In the event the supplier is in error, the purchase order serves as verification of the original order. Additionally, if prices are included on the purchase order form, the designer is quickly able to justify and post the invoice incurred by the purchase order.

As with job numbers and invoice numbers, purchase orders start with any number, and then continue in numerical order. Whether the designer uses preprinted purchase order forms or creates electronic copies, it is helpful, but not necessary, to have forms with preprinted numbers. It is, however, essential to have copies, in hard or digital format. At the very least, the designer will need to send the original copy to the vendor and keep a copy. Studios within large companies have to sync their purchase order system with the bookkeeping departments of the company or have the accounting department issue them, in which case the design studio should be provided a copy.

Filling in the Form

Fill in the number of the purchase order, date of the order, the client and project names, and the project number. Fill in the name of the vendor, the address, and the name of the contact or sales representative at the vendor's place of business, and all contact information. In the space for "schedule," fill in the date the job must be received in the designer's hands and check the appropriate level of urgency. "Overtime" refers to extended business hours, weekends, and holidays. In the space under "specifications," fill in the item number as it pertains to this list, not to any other list or number—always start with "1" and continue numerically down the list. In the space for "description," fill in either the name of the item or instructions for the work to be done or the goods to be delivered. Give quantities, sizes, dimensions, and any other specific information necessary to communicate exactly what is expected. If a drawing is attached, make reference to that information here. If ordering from a catalog, give item numbers, catalog page numbers, descriptions, and so on. Be absolutely precise. Fill in the shipping address and where the goods are to be delivered. The designer should compute the subtotal and total, if costs are known in advance. In the space for "bill to," fill in the name and address of where the supplier is to send the invoice if other than the designer. Print or clearly sign the order and fill in the telephone and extension number where the designer can be reached for questions. Purchase orders should always be signed by the person authorized to issue them, and include direct voice or fax contact information so that the vendor can quickly get in touch with questions or concerns.

Using the Form

A purchase order is a legally binding document. A vendor is entitled to full payment on an item delivered that conforms to the specifications listed in the purchase order, even if the designer has made an error in the specifications. Further, if a purchase order is canceled after the vendor has begun work on an item, even if the cancellation is fairly early, the vendor is still entitled to at least partial payment, if not full—depending upon the terms of the purchase negotiated with the vendor. Purchase order forms should be on letterhead or some other form of stationery that clearly shows the name of the designer, the company name, address, and telephone number.

Remember that both rush and overtime will incur additional costs. It is best to check what these will be in advance. Markups for rush and overtime orders can be as much as 100 percent to 200 percent and more. Overtime refers to work that must be completed overnight, on weekends, or on holidays.

Provide drawings, sketches, samples, as appropriate, to convey a full understanding of the order. The more precise and descriptive an order, the less chance there is of having to accept and pay for mistakes.

Most suppliers prefer to bill the designer directly and usually are not pleased to have to bill a third party. This scenario should be worked out in advance. The supplier will often require a deposit. In many instances, suppliers will also require bank and business references before committing to a large order.

If the designer uses purchase order forms to contract freelance designers to create original designs, it is important to remember to include information pertaining to copyright, future uses, and what, if any, rights the designer has to show the work in portfolio and marketing materials. It is also necessary to obtain the freelancer's social security or federal identification number, so that Form 1099 can be completed (as required by the Internal Revenue Service for any payments of $600 or more to independent contractors).

Purchase Order

Job Number _____

Vendor _____

Address _____

Contact Name _____

Phone _____

Fax _____

E-Mail _____

P.O. Number _____

Date _____

Schedule

Delivery Due _____ [] Regular [] Rush [] Overtime (holiday, weekend, etc.)

Specifications

Item Number	Description	Quantity	Unit Price	Other	Total
1					
2					
3					
4					
5					
6					
7					
8					
9					
10					

Subtotal	$
Shipping/Handling	
Tax	
Total	$
Deposit	
Balance Due	$

Notes _____
_____
_____
_____
_____

Ship to: _____
_____

Bill to: _____
_____
_____

Authorization

Signature _____

Print Name _____

Phone _____

Fax _____

Payables Index

The payables index is used to track all incoming invoices, whether or not they are related to billable jobs. It is handy for checking monthly statements and determining whether or not bills have been approved and posted to job sheets (Form 2). It also contains all of the necessary information to relocate lost bills. Incoming invoices should be gathered on a daily or biweekly basis (depending upon the volume of incoming bills). After recording the required information to the payables index, distribute the bills to those who will be approving them. (A closed-sided manila file folder with each person's name on the tab is handy for this purpose.) This is assuming, of course, that purchasing and placing orders is not a centralized function in the operation. Some firms invest design teams with the responsibility of placing and tracking orders; other firms prefer to have a purchasing agent handle all of these kinds of needs. Either way, there should be a system to check purchases during each step of the process: first, when placing the order; then to check the delivery against the order specifications; and finally to examine the invoice against both the purchase order and the condition of delivery, if any adjustment is required.

Filling in the Form

In the first column, fill in the day the bill was received, the name of the vendor or supplier, the invoice number, its date, and the total amount due. Under "attention" list the person in the firm who is responsible for approving this particular bill. Write the date it was returned in the "approved" column. Job-related bills are now ready for posting to job sheets, after which they can be paid and filed; bills not related to jobs can be paid and filed directly.

Using the Form

All bills for expenses incurred by a specific job, or for general supplies for the firm, should be checked by the individual who ordered the corresponding material or services. (Generally, project managers on large assignments have the responsibility of reviewing such invoices.)

Whoever is responsible for approving invoices should indicate the following on each one:

❏ The project job number (if it was an expense incurred for the project, but not a billable item, indicate as such with an "NB" next to the job number).

❏ If the bill is for a studio expense and not related to any specific job, simply indicate "studio"—or provide the department number.

❏ If the bill is for more than one job, divide the sum appropriately and indicate the amounts applicable to the separate jobs.

❏ The person's initials and the date.

If the bill is incorrect, this is the time to make adjustments with the vendor. Ask the vendor to reissue a correct invoice, if possible, rather than making the corrections by hand.

File a copy of all invoices billable to projects in file folders (or three-ring binders) identified on the outside by individual job names and numbers. Every job should have this "backup" file to collect copies of all reimbursable expenses (which are then sent to the client when the job is billed to substantiate the listed expenses).

Payables Index

Date Received	Vendor/Supplier	Amount	Invoice Number	Amount Due	Attention	Approved	Posted

Billing Index

The billing index is the list of all monies billed to clients for ongoing or completed work. Knowing the amount and timing of expected revenue is critical information in any business.

Bookkeepers and accountants are familiar with "accounts receivables," and among several other types of ledgers, they keep a separate ledger of all outgoing invoices. It is helpful on the studio level to be able to quickly and easily check when jobs have been billed, whether or not they have been paid, and how long bills have been outstanding (not paid).

The billing index is also the source of all invoice numbers. Invoice numbers are to billing what job numbers are to cost accounting. An invoice number is an identifying "tag" that serves to keep track of invoices within both the designer's and the client's bookkeeping systems. When assigning invoice numbers, begin with any number and then follow consecutively thereafter. It is best not to link invoice numbers to job numbers or purchase order numbers. An uncomplicated, independent, sequential list of numbers is less likely to create confusion for both manual and computerized billing systems. More information about creating numbering systems is given in the job index (Form 1).

Filling in the Form

Fill in the date that will appear on the invoice under "date." Fill in the invoice number and the job number of the assignment being billed. For "billed to" fill in the name of the client as it appears on the job sheet. Fill in the job name, as designated on the job index. Fill in the fees, the amount of reimbursable expenses, and tax, if applicable, in the designated columns. You might want to separate travel from marked-up reimbursable items, in which case, use the "other" column and rename it "travel." A formula has been built into the spreadsheet to tally the total amount. The "paid" columns are for the date and check number when the payment is received.

Using the Form

The billing index is used to generate invoices for fees and costs. Projects might be billed on a flat fee basis, in which case billing is phased over the course of the project. It is customary to collect a 50 percent deposit upon signing of the agreement, 25 percent upon completion of concept development, and 25 percent upon completion of design drawings and specifications. For long-term projects and those with very large fees, phased billing may be negotiated with 1/3 to start, and a fixed amount to be billed monthly, along with out-of-pocket expenses thereafter.

Use this index together with the project's job sheet and work change orders (a summary of which should be on the job sheet) when preparing invoices. First, determine the amount to be billed, based upon the terms of the arrangement in place; tally up all the reimbursables; and assign a number to the invoice from the billing index. Write up the invoice and fill in the rest of the information indicated on the billing index. On the job sheet, jot down the invoice number, date and amount. Use the work change order number on the invoice to reference any changes to the expected billing.

Billing Index

Date	Invoice #	Job Number	Billed To	Job Name	Fee	Expenses	Other	Tax	Total	Date Paid	Check No.

Invoices

Billing is the financial pipeline of every design studio. Considering that turnaround time for payment can be anywhere from thirty to ninety days and more, it is imperative that billing be done quickly and regularly. In every project agreement and contract, even for add-on work, establish a payment schedule, even for small assignments. At the least, arrange for an advance payment against the total fee with the balance and reimbursable expenses to be due upon completion of the work. If the job is based on a one-time final payment, it should be billed immediately upon completion. More information about payment schedules is available in the billing index (Form 20).

Two different sample invoice forms are shown here. Although the total billing information contained may be identical, the second form is much more detailed than the first. It is better not to have forms that have preprinted numbers. Computer-generated forms allow for easy customization and outputting as needed. The billing index (Form 20) is the running log and source of all invoice numbers.

Compensation Structures

Several different compensation models have been discussed throughout this book. To summarize:

Flat Fee

A flat fee is the overall fee for the project, as determined and specified in the project contract. Billing may be phased by either percentages of work/phases completed or by regular time intervals (e.g., monthly, quarterly, etc.). Depending upon the project contract, the flat fee might be only for industrial design services, with reimbursable items listed separately, or a project may have one entire flat fee that includes all reimbursables. These points are negotiable in advance and it is critically important that the designer understands very clearly what the scope of work will entail, in order to accurately estimate a flat fee. There should be written caveats in all flat fee contracts that protect the designer in the event the scope of work changes, or the schedule slips as a result of client-caused issues.

Hourly Fee

In this situation, billing is based on hourly fees for time actually worked. The hourly (or weekly) fee may be per individual rates, or on a tiered rate for different levels or types of personnel. Time sheets or summary spreadsheets of time expended should be available for clients to see, if they so wish, at regular intervals or upon billing. Reimbursable expenses are included in these invoices, but listed separately.

Hourly Fee with a Cap (Pre-set Maximum)

Billing based on an hourly rate structure (as above), but with a specified maximum limit. This structure usually has predetermined rates and a specified maximum number of hours or maximum amount for the total fee. It is important to include a caveat in the project contract that allows for additional billing dependent upon various contingencies, such as work change orders.

Royalty

Designers sometimes work with companies on a royalty basis. The arrangement may or may not include the designer's cost to develop the product—depending upon the level of investment and return the designer is willing to risk. These arrangements are potentially profitable in the long run; the prospect of collecting "passive" income for years on a product that is doing well is very attractive. However, as in all business ventures, there is a risk that very little or nothing may come of the endeavor. These are factors that must be weighed carefully in advance, and with the help of an experienced attorney who will structure the deal so that the maximum benefit is reaped by the designer should a product succeed.

Filling in the Form: Invoice 1

Use this form for billing simple fees and/or a limited number of reimbursable items. Fill in the date of the invoice, its number (from the billing index), and the project number (from the job index). Fill in the complete billing "to" information, including company name, if applicable, and the billing address. "Attention" is usually used with corporate or institutional entities, and might refer either to the project contact, the purchasing agent, or simply "accounts payable." Under "description," if a purchase order has been issued by the client, fill in that number. State the name of the assignment next to "project name." If the billing is in response to, or affected by, a work change order(s), fill in that number or numbers.

The easiest way to write the "description" for an invoice is to copy or summarize with a list in the same language used in the assignment's proposal, letter of agreement, or contract. If the agreement contains a schedule of payment, the invoice can be a direct copy of that schedule.

If the invoice is based on an overall flat fee, list the services performed, but do not assign dollar amounts to each service, just summarize the list with an all-encompassing phrase, such as "concept development fee." Reimbursable items should be grouped by type, with one total number for each. If this form is used to bill staff time, use either the individual person's name and/or title, and the total dollar amount of their work under "fees." For extensive, detailed lists, use invoice 2. Subtotal all the items, indicate the amount of tax, if applicable, remember to indicate and deduct from the subtotal, any pre-paid fees or retainers that are applicable to this invoice and fill in the total for the entire bill. In

the "terms" statement, indicate if the invoice is payable "upon receipt" or whatever the time-frame that has been agreed to with the client. Even if clearly noted in the contract, it is advisable to reiterate the terms of copyright and usage rights that the designer wishes to retain.

Filling in the Form: Invoice 2

Use invoice 2 for billing that is based upon a variety of factors, such as percentage fees, hourly rates for personnel, and reimbursable items. Fill in the complete billing "to" information, including company name, if applicable, and the billing address. "Attention" is usually used with corporate or institutional entities, and might refer to either the project contact, the purchasing agent, or simply "accounts payable." If a purchase order has been issued by the client, fill in that number. State the name of the assignment next to "project name." If the billing is in response to, or affected by, a work change order(s), fill in that number or numbers.

Check the project phase being billed, if applicable. Include any relevant detail under "notes," and include the fee amount next to "fee subtotal." To bill the cost of personnel, list either names and titles (Spencer B., Sr. designer), or only titles (junior designer), or tiered task descriptions (executive, senior, junior, etc.), indicate the hours, the hourly rates and the total for each line. Tally the personnel costs next to "personnel subtotal." The "reimbursable expenses" included in this form are suggested items. The list may be inapplicable to your purposes, or it may be incomplete. Feel free to adjust these items as needed. Include any relevant notes, if needed under "notes." If you are passing through vendor invoices, treat them in their entirety do not break them out for the client in your invoice. You can or should include these invoices as backup, depending upon your arrangement with the client on this issue. For example, if a model maker sends you a bill that itemizes all the materials used, only fill in the total amount of the invoice and attach the invoices if backup scrutiny is expected. For the complete invoice total, subtotal the reimbursable items, add up the fees, personnel costs and reimbursables. Be sure to indicate, and deduct from the subtotal, any paid deposits or retainers that are applicable to this invoice.

Using the Form(s)

One of the inevitable snags in producing bills quickly is that suppliers and freelancers often do not submit their bills in time to be approved and posted to project job sheets. This results in incomplete backup information. First, it is necessary to urge all outside providers to be timely in their billing, and second billing may be broken out, separating fees from costs. An invoice for the fee or billable time and all available costs may be billed with a note on the invoice, clearly written in an obvious location, stating, "Additional production/reimbursable costs to follow." When all such costs have been received, approved and posted, a second invoice can follow, with a notation under "description":

Additional production costs, as per invoice #____, dated ____.

For pre-arranged markups on hourly rates and/or reimbursable items, the invoice should automatically reflect the marked-up rates. It is not necessary to show the math on the invoice itself. While it is not necessary to itemize every markup computation on reimbursable items, it is recommended that the backup bills for each grouping of items be stapled together with an adding machine tape showing both the tally of the attached bills and the markup computation, or with a summary spreadsheet cover. This is a courtesy to the client's bookkeeper and should help move your invoices through your client's accounting department more efficiently.

Invoice 1

Company Name _____   Date _____

Address _____   Invoice Number _____

_____

_____   Project Number _____

Attention _____

Project Name _____

Purchase Order No. _____

Work Change Order No. _____

Description	Fees	Costs	Total
Subtotals	$ -	$ -	$ -
		Tax	
		Total	$ -

Terms: Invoices are payable upon receipt.

Design documents and materials including, but not limited to, drawings, renderings, models, prototypes, specifications, sample boards, and all other design documents and materials are the exclusive property of the designer. Exclusive copyright of these materials is reserved by the designer. Upon full payment of all fees and costs, the client is granted the right to use the designs contained in these materials as specified in project contract only. All other rights remain the exclusive property of the designer.

Invoice 2

Company Name _____

Address _____

_____

_____

_____

Attention _____

Date _____

Invoice Number _____

Project Number _____

Purchase Order No. _____

Work Change Order No. _____

Project Name _____

Project Phase		Fee Amount
❏ Strategic Analysis and Planning		
❏ Creative Development		
❏ Design Development		
❏ Implementation		
	Total	$ -
Other/Notes		Fee Amount
	Fee Subtotal	$ -

Personnel	Hours	Rate	
Personnel Subtotal			$ -

Reimbursable Expenses	Notes	Total
Color Copies		
Color Transfers/Other Color Processes		
Computer Prints		
Courier Services/Shipping		
Fabrication		
Illustration/Other Rendering		
Legal Fees		
Local Messengers		
Modelmaking		
Photographic & A/V Supplies		
Photographic Prints		

Photography		
Portable Media (Disks, CDs, etc.)		
Presentation Materials		
Prototypes		
Reference Materials		
Reproduction		
Research & Database Services		
Research Materials		
Scanning		
Shipping		
Special Supplies/Materials		
Telecommunications		
Travel		
Airfare		
Ground Transportation		
Lodging		
Meals		
Entertainment		
Other		
	Reimbursable Expenses Subtotal	$ -
	Total Due	$ -

Terms: Invoices are payable upon receipt.

Design documents and materials including, but not limited to, drawings, renderings, models, prototypes, specifications, sample boards, and all other design documents and materials are the exclusive property of the designer. Exclusive copyright of these materials is reserved by the designer. Upon full payment of all fees and costs, the client is granted the right to use the designs contained in these materials as specified in project contract only. All other rights remain the exclusive property of the designer.

Statements and Final Notice

Clients may be instructed to pay invoices within any reasonable amount of time, but in reality it is more likely that clients will pay according to their own payment cycles. However, once payment for an invoice is overdue (anywhere from ten to thirty days, depending upon the specific time frame established in the proposal or agreement for the assignment), the designer has two options. To call the client, mention the possibility of an oversight, and ask the client, or client contact to look into the matter. And/or, the designer can send the client a billing statement, such as Form 23. Such statements are routinely sent by all vendors and creditors, usually monthly, and are merely a summary of the amounts due for payment.

In the unfortunate event that there is no response to the first statement or call, another call can be placed to the client or the designated contact. If this does not yield satisfaction within a very short time, a second notice, such as the one on page 96, should be sent. This notice is a virtual copy of the statement, with slightly different language. Some designers simply stamp a copy of the original invoice with the statement "second notice," or "overdue," but with this method, the markup on overdue invoices is disregarded.

By the time of the final notice, Form 24, the statement should be sent in a way that the client has to sign a receipt acknowledging its delivery: messenger, telegram, registered mail, UPS, FedEx, or some other couriers service would serve this purpose.

Filling in the Form

Fill in the date, the name of the client (individual or corporate), the address, and the name of the contact on the assignment or the head of accounts payable next to "attention."

Fill in the name of the project, its job number and the client contact, if different from the person receiving the invoice. If several assignments are involved, state the overall name of the account (this may be the client's name or the name of the corporate entity). List the unpaid invoices. Calculate the amount of the service charge and fill in the name and contact information of the person whom the client should contact in the designer's office. For Form 24, final notice, list the overdue invoice number(s), date(s), and total amount(s).

Using the Form

There is no guaranteed way of avoiding the problem of collecting payment on late bills. It is critically important to have all of the financial details of a project detailed in a clearly written letter of agreement or contract. It is very important for the designer to have a copy of such a document with the client's signature. It is also necessary to bill projects promptly and regularly.

Establishing advance payments in the form of retainers and deposits and a schedule of payments is also extremely useful. It is better to spot payment problems while the designer still has some leverage, or hasn't invested too much into the project.

The last recourse is to turn the account over for collection either by a reputable agency or attorney. Fees for this service range from 25 to 40 percent (and more) of the monies collected. Keep in mind that collection agencies can only ask for money. They are not licensed to practice law and cannot bring lawsuits. If the client is unlikely to pay, retaining an attorney at an hourly fee might be the best approach.

For amounts of less than a few thousand dollars, depending upon locale, small claims court may be a viable option. Local court offices provide information about filing claims.

Tenacity is important to the success of collecting, but in some cases it may be necessary to be flexible in accepting partial payment or a revised payment schedule.

Second Notice   Date _____

To _____
Attention _____
Reference _____

_____

Please be advised that payment for the following have not been received as of the date of this statement.

Invoice Date _____
Invoice No. _____
Invoice Amount _____
Service Charge _____

Total Due _____

This is the second notice you have received about this overdue account. If you cannot make immediate payment in the full amounts shown, please contact us right away.

Contact _____   Telephone Number _____
_____   Fax Number _____
_____   E-mail _____

Statement

Date _____ .

To _____

Address _____

_____

_____

Attention _____

_____

Project Name _____

Project Number _____

Project Contact _____

Please be advised that payment for the following has not been received as of the date of this statement.

Date of Invoice	Invoice Number	Invoice Amount Due	Service Charge	Total Due
_____	_____	_____	_____	_____
_____	_____	_____	_____	_____
_____	_____	_____	_____	_____
_____	_____	_____	_____	_____
_____	_____	_____	_____	_____
			Total Due	$ _____

Your prompt attention and earliest payment would be greatly appreciated. Please contact us if you have questions about this statement. Please note that invoices not paid according to terms are subject to 1.5 percent monthly service charge.

Contact _____

Telephone Number _____

Fax Number _____

E-mail _____

Design documents and materials including, but not limited to, drawings, renderings, models, prototypes, specifications, sample boards, and all other design documents and materials are the exclusive property of designer. Exclusive copyright of these materials is reserved by the designer. Upon full payment of all fees and costs, client is granted the right to use the designs contained in these materials as specified in project contract only. All other rights remain the exclusive property of the designer.

Final Notice

Date _____

Name of Company _____

Address _____

_____

_____

Attention _____

Invoice Number	Invoice Date	Invoice Amount
	Total	$

This account is now in arrears. We have repeatedly requested payment and have neither received payment nor have we been contacted with an explanation.

We must collect immediately, and, if payment is not received within ____ days of the date of this notice, we have no choice but to turn this account over for collection. Be aware that this process may result in additional legal and court costs to you and may damage your credit rating.

It is not too late to contact us:

Contact _____

Telephone Number _____

Fax Number _____

E-mail _____

Marketing Call Log and Qualifying Checklist

Logic would dictate that talent, experience, sensible business practices, and the ability to manage people and endless project details are all key prerequisites to a successful industrial design practice. In fact, these are useful attributes, but without the constant pursuit and flow of new projects, no independent firm can stay in business. Even an in-house industrial design studio has to maintain a level of productivity that justifies its existence within a larger organization. Every industrial design firm should have a realistic and well-thought-out plan to promote its capabilities. There have to be funds for some reasonable expenditure on marketing materials including press releases, brochures, photography, and a Web site. It also requires the investment of time by both principals and staff members.

Firms that are run by two or more partners are usually at an advantage in this area, because it is very likely that one of the partners is well suited to the marketing and promotion needs of the business. It is also possible that a sole proprietor is someone with a strong vision and force of personality who is able to drive the business in directions that will reap new sources of work. Given the advantage of personality, vision, and drive, with a plan and regular procedures, the chances for long-term success are much greater. If there is no one within the firm who can devote the time and resources needed to develop such a plan, it may be necessary to hire a marketing consultant. Be sure to select someone who either understands your business very well or has extensive experience and proven success in those markets in which you wish to develop a presence.

With all the promotion tactics and techniques that are available (see "Using the Form" on page 100), the most immediate and most valuable source of new work can be found in your client list—past and present. Word-of-mouth, personal introductions, and references are the cheapest and most reliable sources of new business that exist. There is no more credible recommendation than from a former or current client. Endeavor to understand their design needs and challenges, and help them realize their vision by delivering the best quality possible for the price. Do all you can to mitigate problems and be flexible and creative in solving them. Do all you can to make your clients feel that they have received the best possible services that you can offer. The long-term viability of a design firm is based upon relationships, not on individual projects. It's not easy to build loyalty at a time when clients are seeking the lowest fees; it requires consistency both of product excellence and account management. The client has to sense that the designer genuinely cares about not only the product, but about the client's needs and concerns.

Filling in the Forms

Part one of this form is the call log. This form can be copied and repeated endlessly as a separate log, or it can be attached to part two, the qualifying checklist. The qualifying checklist can stand on its own, or as noted, be attached to the individual call log that generated the need for further information.

Call Log

Fill in the date of the call, and who made the call (next to "by"). Jot down the name and title of the contact that was reached and any comments that may be useful to follow up. Under "results/next steps" check the appropriate box and the recommended date for the action to be taken. Make sure to follow through either personally or transmit the needed information to the person who is supposed to follow up.

Qualifying Checklist

Fill in the date the form is being generated and who is filling in the form (next to "by"). Fill in all of the information about the prospective client, including company and contact name, address, and all contact information such as phone, cell and fax numbers as well as e-mail address. For project information, fill in the general name of the industry that the product is in, such as technology, health and medicine, hospitality, and so on, under "industry," identify the product or element, and how many will be required (if known). Indicate the potential fee and under "schedule," in which quarter the project might start, and/or when revenues will be earned. Indicate any known history of the client and list possible competitors. Include any special remarks and other information under "notes." Finally, check off the applicable "results/next steps," and the dates they should be taken.

Using the Form

Here is a brief outline of the steps, questions and actions involved in developing a marketing plan:

SELF-ASSESSMENT

❏ Mission: Why are we in business? What do we hope to accomplish? What do we want to be known for?

❏ Strengths: What kinds of special skills and unique experiences do we have? Are we in a highly specialized niche?

❏ Experience: Develop a case history sheet for every significant project, including beauty shots of the finished products

❏ Capabilities: What is the range of skills we have? At what relative levels?

❏ Capacity: How much work can we do well at one time?.

❏ Client list: Develop a client list that has the dates and names and notes for all projects worked on with each client

❏ Competitors: Who are the closest competitors in our market? How do they promote themselves?

MARKET-ASSESSMENT

❏ Using resources such as newspapers, trade and government journals, local newsletters, industry reference guides and trade association resources, study possible sources of work in:

❏ Niche markets—specific trades and industries
❏ Local—look for areas of growth
❏ Regional—look for areas that are underserved

ACTION PLANS

❏ Identity and Visibility

❏ Develop logo, stationery, business cards, brochures, signage, leave-behinds, Web site, and other image materials that give a clear message of the firm's professionalism and expertise
❏ Enter completed work in competitions
❏ Enter showcase events—trade shows and other opportunities to raise visibility in the trades and local industry
❏ Appear/speak at lectures, industry and trade events
❏ Serve on design juries
❏ Write articles, send pictures to industry newsletters, and consumer, trade and industry magazines
❏ Send press releases to announce special events
❏ Hire a public relations firm
❏ Consider paid advertising

❏ Lead Generation

❏ Former clients—reference list and quotes
❏ Industry and trade contacts—reference list and quotes
❏ Develop/buy call and mail lists—cull qualified leads into a database

❏ Institute regular system of cold calling and follow-up

❏ Institute regular mailings and follow-up

❏ Lead Qualification and Pursuit

   ❏ Cull warm leads and institute process for follow through

❏ Have boilerplates ready of initial estimates and proposals for quick response to RFPs (requests for proposal)

❏ Develop presentation materials and skills

❏ Selling and closing—study the art of persuasion

Marketing Call Log

Date _____ By _____
Contact: _____ Phone _____
Name _____
Title _____
Notes _____
_____
_____
_____
_____

Results/Next Steps:

❑ No Lead   ❑ Will call   ❑ Call back   ❑ Research   ❑ Send materials   ❑ Appointment
      Date _____   Date _____   Date _____   Date _____   Date _____   Date _____

Date _____ By _____
Contact: _____ Phone _____
Name _____
Title _____
Notes _____
_____
_____

Results/Next Steps:

❑ No Lead   ❑ Will call   ❑ Call back   ❑ Research   ❑ Send materials   ❑ Appointment
      Date _____   Date _____   Date _____   Date _____   Date _____   Date _____

Date _____ By _____
Contact: _____ Phone _____
Name _____
Title _____
Notes _____
_____
_____

Results/Next Steps:

❑ No Lead   ❑ Will call   ❑ Call back   ❑ Research   ❑ Send materials   ❑ Appointment
      Date _____   Date _____   Date _____   Date _____   Date _____   Date _____

Date _____ By _____
Contact: _____ Phone _____
Name _____
Title _____
Notes _____
_____
_____

Results/Next Steps:

❑ No Lead   ❑ Will call   ❑ Call back   ❑ Research   ❑ Send materials   ❑ Appointment
      Date _____   Date _____   Date _____   Date _____   Date _____   Date _____

Qualifying Checklist

Date _____   By _____

PROSPECT INFORMATION

Company _____   Phone _____

Contact Name _____   Cell _____

Address _____   Fax _____

                                           Email _____

PROJECT

Industry _____   ❏ New Design

Product _____   ❏ Adaptation

Quantity _____   ❏ Design Only

Potential Fee _____   ❏ Design and Implementation

Schedule _____

      Q1   Q2   Q3   Q4

CLIENT HISTORY   COMPETITION

       ❏ No prior contact   _____

       ❏ Aware of firm   _____

       ❏ Knows firm well   _____

       ❏ Other   _____

NOTES _____

_____

_____

_____

Results/Next Steps

❏ Call back   ❏ Send materials   ❏ Research   ❏ Send letter   ❏ Proposal   ❏ Appointment

Date _____   Date _____   Date _____   Date _____   Date _____   Date _____

Commercial Lease
Sublease
Lease Assignment

Every business, whether small or large, must find suitable space for its activities. For the sole proprietor, the studio may be in the home. For a larger enterprise, the studio or office is likely to be in an office building. While some businesses may own their offices, most will rent space from a landlord. The terms of the rental lease may benefit or harm the business. Large sums of money are likely to be paid not only for rent, but also for security, escalators, and other charges. In addition, the tenant is likely to spend additional money to customize the space by building in offices, darkrooms, reception areas, and the like. Form 27 and the accompanying analysis of how to negotiate a lease are designed to protect the tenant from a variety of troublesome issues that may arise.

The old saw about real estate is, "Location, location, location." Location builds value, but the intended use of the rented premises must be legal if value is to accrue. For example, sole proprietors often work from home. In many places, zoning laws govern the uses that can be made of property. It may be that an office at home violates zoning restrictions against commercial activity. What is fine in one town—a studio in an extra room, for example—may violate the zoning in the next town. Before renting or buying a residential property with the intention of also doing business there, it's important to check with an attorney and find out whether the business activity will be legal.

In fact, it's a good idea to retain a knowledgeable real estate attorney to negotiate a lease, especially if the rent is substantial and the term is long. That attorney can also give advice as to questions of legality. For example, what if the premises are in a commercial zone, but the entrepreneur wants to live and work there? This can be illegal and raise the specter of eviction by either the landlord or the local authorities.

In addition, the lease contains what is called the "use" clause. This specifies the purpose for which the premises are being rented. Often the lease will state that the premises can be used for the specified purpose and no other use is permitted. This limitation has to be observed or the tenant will run the risk of losing the premises. The tenant therefore has to seek the widest possible scope of use, certainly a scope sufficient to permit all of the intended business activities.

Another risk in terms of legality involves waste products or emissions that may not be legal in certain areas. Again, an attorney should be able to give advice about whether a planned activity might violate environmental protection as well as zoning laws.

The premises must lend themselves to the intended use. Loading ramps and elevators must be large enough to accommodate whatever must be moved in to set up the office (unless the tenant is willing and able to use a crane to move large pieces of equipment in through windows) and any products that need to be shipped once business is underway. Electric service must be adequate for air conditioning, computers, and other machinery. It must be possible to obtain telephone service and high-speed Internet connections. If the building is commercial and the tenant intends to work on weekends and evenings, then it will necessary to ascertain whether the building is open and heated on weekends and evenings. If clients may come during off hours, it is even more important to make sure the building is open and elevator service is available.

Who will pay for bringing in electric lines, installing air conditioners, building any needed offices, installing fixtures, painting, or making other improvements? This will depend on the rental market, which dictates the relative bargaining strengths of the landlord and tenant. The lease (or an attached rider) must specify who will pay for each such improvement. It must also specify that the landlord gives the tenant permission to build what the tenant needs.

A related issue is who will own structures and equipment affixed to the premises when the lease term ends. Since the standard lease gives ownership to the landlord of anything affixed to the premises, the tenant must have this provision amended if anything of this nature is to be removed from the premises or sold to a new incoming tenant at the end of the lease term.

If the tenant has to make a significant investment in the costs of moving into and improving the premises, the tenant will want to be able to stay long enough to amortize this investment. One way to do this is to seek a long lease term, such as ten years. However, the landlord will inevitably want the rent to increase during the lease term. This leads to a dilemma for the tenant who can't be certain about the needs of the business or the rental market so far into the future.

One approach is to seek the longest possible lease, but negotiate for a right to terminate the lease. Another strategy would be to seek a shorter initial term, but have options to extend the lease for additional terms at agreed upon rents. So, instead of asking for a ten-year lease, the tenant might ask for a four-year lease with two options for additional extensions of three years each.

Yet another tactic, and probably a wise provision to insist on in any event, is to have a right to sublet all or part of the premises or to assign the lease. The lease typically forbids this, so the tenant will have to demand such rights. Having the ability to sublet or assign offers another way to take a long lease while keeping the option to exit the premises if more space is needed, the rent becomes too high, or other circumstances necessitate a move. However, the right to sublet or assign will be of little use if the real estate market turns down, so it should probably be a supplement to a right to terminate or an option to renew.

Part of the problem with making a long-term commitment is that the stated rent is likely not going to be all of the rent. In other words, the lease may say that the rent is $2,400 per month for the first two years and then escalates to $3,000 per month for the next two years. But the tenant may have to pay for other increases or services. It is common for leases to provide for escalators based on increases in real estate taxes, increases in inflation, or increases in indexes designed to measure labor or operating costs. In fact, the standard lease will make the tenant responsible for all nonstructural repairs that are needed. The tenant has to evaluate all of these charges not only for reasonableness but for the impact on what will truly be paid each month. There may also be charges for refuse removal, cleaning (if the landlord provides this), window washing, water, and other services or supplies. If the landlord has installed submeters for electricity, this may result in paying far more to the landlord than would have been paid to the utility company. It may be possible to lessen the markup, obtain direct metering, or, at the least, factor this cost into considerations for the budget.

Faced with costs that may increase, the tenant should try to determine what is realistically likely to occur. Then, as a safeguard, the tenant might ask for ceilings on the amounts that can be charged in the various potential cost categories.

Leases are usually written documents. Whenever a written agreement exists between two parties, all amendments and modifications should also be in writing and signed by both parties. For example, the lease will probably require one or two months' rent as security. The tenant will want this rent to be kept in an interest bearing account in the tenant's name. But if the parties agree to use part of the security to pay the rent at some point, this should be documented in a signed, written agreement.

The tenant should also seek to avoid having personal liability for the lease. Of course, if the tenant is doing business as a sole proprietor, the tenant by signing the lease will assume personal liability for payments. But if the tenant is a corporation, it would certainly be best not to have any officers or owners give personal guarantees that would place their personal assets at risk.

Leases can grow to become thick sheaves of paper filled with innumerable clauses of legalese. Since the lease agreement can be such an important one for a business, it is ideal to have a knowledgeable attorney as a companion when entering its maze of provisions. If an attorney is too expensive and the lease is short-term and at an affordable rent, this discussion may give a clue that will help the tenant emerge from lease negotiations with success.

Form 27 is an educational tool. It has been drafted more favorably to the tenant than would usually be the case. Because it would be unlikely for the tenant to present a lease to the landlord, Form 27 shows what the tenant might hope to obtain from the negotiations that begin with a lease form presented to the tenant by the landlord.

Forms 28 and 29 relate to a transformation of the role of the tenant. Whether to move to superior space, lessen cash outflow, or make a profit, the tenant may want to assign the lease or sublet all or a portion of the premises. Form 28 is for a sublease in which the tenant essentially becomes a landlord to a subtenant. Here the tenant must negotiate the same issues that were negotiated with the landlord, but from the other point of view. So, for example, while a corporate tenant would resist having its officers or owners give personal guarantees for a lease, when subletting the same corporate tenant might demand such personal guarantees from a corporate subtenant. Form 29 for the assignment of a lease is less complicated than the sublease form. It essentially replaces the tenant with an assignee who becomes the new tenant and is fully responsible to comply with the lease. A key issue in such an assignment is whether the original tenant will remain liable to the landlord if the assignee fails to perform. A related issue with a corporate assignee would be whether the assignee would give a personal guarantee of its performance. In any event, both the sublease and the assignment will usually require the written consent of the landlord. This consent can take the form of a letter signed by both the tenant and the landlord. In the case of an assignment, the letter of consent could also clarify whether the original tenant would have continuing liability pursuant to the lease.

Filling in the Form: Form 27

In the preamble fill in the date and the names and addresses of the parties. In paragraph 1 give the suite number, the approximate square footage, and floor for the premises as well as the address for the building. In paragraph 2 give the number of years as well as the commencement and termination dates for the lease. In paragraph 3 give a renewal term, if any, and indicate what modifications of the lease would take effect for the renewal term. In paragraph 5 indicate the annual rent. In paragraph 6 specify the number of months' rent that will be given as a security deposit. In paragraph 7 indicate any work to be performed by the landlord and the completion date for that work. In paragraph 8 specify the use that the Tenant will make of the premises. In paragraph 9 give the details as to alterations and installations to be done by the Tenant and also indicate ownership and right to remove or sell these alterations and installations at the termination of the term.

In paragraph 10 indicate any repairs for which the tenant shall be responsible. In paragraph 13 check the appropriate box with respect to air conditioning, fill in the blanks if the tenant is not paying for electricity or the landlord is not paying for water, and indicate who will be responsible for the cost of refuse removal. In paragraph 14 state whether or not the landlord will have a key to provide it access to the premises. In paragraph 15 indicate the amount of liability insurance the tenant will be expected to carry as well as any limitation in terms of the landlord's liability for the tenant's losses due to fire or other casualty affecting the building. In paragraph 18 indicate which state's laws shall govern the agreement. Have both parties sign and append a rider, if necessary. Leases frequently have riders, which are attachments to the lease. This gives the space to add additional provisions or amplify details that require more space such as how construction or installations will be done. If there is such a rider, it should be signed by both parties.

Negotiation Checklist for Form 27

❏ If the tenant is a corporation, do not agree that any owner or officer of the corporation will have personal liability with respect to the lease.

❏ Consider whether, in view of the rental market, it may be possible to negotiate for a number of months rent-free at the start of the lease (in part, perhaps, to allow time for construction to be done by the tenant).

❏ In addition to rent-free months, determine what construction must be done and whether the landlord will do the construction or pay for all or part of the construction. (Paragraph 7)

❏ Since landlords generally rent based on gross square footage (which may be 15–20 percent more than net or usable square footage), carefully measure the net square footage to make certain it is adequate. (Paragraph 1)

❏ Specify the location and approximate square footage of the rented premises as well as the location of the building. (Paragraph 1)

❏ Indicate the duration of the term, including starting and ending dates. (Paragraph 2)

❏ Determine what will happen, including whether damages will be payable to the tenant if the landlord is unable to deliver possession of the premises at the starting date of the lease.

❏ Especially if the lease gives the tenant the right to terminate, seek the longest possible lease term. (Paragraphs 2 and 4)

❏ Specify that the lease shall become month-to-month after the term expires. (Paragraph 2)

❏ If the landlord has the right to terminate the lease because the building is to be demolished or taken by eminent domain (i.e., acquired by a governmental body), consider whether damages should be payable by the landlord based on the remaining term of the lease.

❏ If the tenant is going to move out, consider whether it would be acceptable for the landlord to show the space prior to the tenant's departure and, if so, for how long prior to the departure and under what circumstances.

❏ Seek an option to renew for a specified term, such as three or five years, or perhaps several options to renew, such as for three years and then for another three years. (Paragraph 3)

❏ Seek the right to terminate the lease on written notice to the landlord. (Paragraph 4)

❏ Although the bankruptcy laws will affect what happens to the lease in the event of the tenant's bankruptcy, do not agree in the lease that the tenant's bankruptcy or insolvency will be grounds for termination of the lease.

❏ Specify the rent and indicate when it should be paid. (Paragraph 5)

❏ Carefully review any proposed increases in rent during the term of the lease. Try to resist adjustments for inflation. In any case, if an inflation index or a porter's wage index is to be used to increase the rent, study the particular index to see if the result is likely to be acceptable in terms of budgeting.

❏ Resist being responsible for additional rent based on a prorated share of increases in real estate taxes and, if such a provision is included, make certain how it is calculated and that it applies only to increases and not the entire tax.

❏ Resist having any of the landlord's cost treated as additional rent.

❏ Indicate the amount of the security deposit and require that it be kept in a separate, interest-bearing account. (Paragraph 6)

❏ Specify when the security deposit will be returned to the tenant. (Paragraph 6)

❏ If the tenant is not going to accept the premises "as is," indicate what work must be performed by the landlord and give a completion date for that work. (Paragraph 7)

❏ If the landlord is to do work, consider what the consequences should be in the event that the landlord is unable to complete the work on time.

❑ Agree to return the premises in broom clean condition and good order except for normal wear and tear. (Paragraph 7)

❑ Seek the widest possible latitude with respect to what type of businesses the tenant may operate in the premises. (Paragraph 8)

❑ Consider asking that "for no other purpose" be stricken from the use clause.

❑ If the use may involve residential as well as business use, determine whether this is legal. If it is, the use clause might be widened to include residential use.

❑ Obtain whatever permissions are needed for the tenant's alterations or installations of equipment in the lease, rather than asking for permission after the lease has been signed. (Paragraph 9)

❑ If the tenant wants to own any installations or equipment affixed to the premises (such as an air conditioning system), this should be specified in the lease. (Paragraph 9)

❑ If the tenant owns certain improvements pursuant to the lease and might sell these to an incoming tenant, the mechanics of such a sale would have to be detailed because the new tenant might not take possession of the premises immediately and the value of what is to be sold will depend to some extent on the nature of the lease negotiated by the new tenant. (Paragraph 9)

❑ Nothing should prevent the tenant from removing its furniture and equipment that is not affixed to the premises. (Paragraph 9)

❑ Make the landlord responsible for repairs in general and limit the responsibility of the tenant to specified types of repairs only. (Paragraph 10)

❑ Obtain the right to sublet all or part of the premises. (Paragraph 11)

❑ Obtain the right to assign the lease. (Paragraph 11)

❑ If the landlord has the right to approve a sublease or assignment, specify the basis on which approval will be determined (such as "good character and sound finances") and indicate that approval may not be withheld unreasonably. (Paragraph 11)

❑ If the tenant has paid for extensive improvements to the premises, a sublessee or assignee might be charged for a "fixture" fee.

❑ If a fixture fee is to be charged, or if the lease is being assigned pursuant to the sale of a business, do not give the landlord the right to share in the proceeds of the fixture fee or sale of the business.

❑ Guarantee the tenant's right to quiet enjoyment of the premises. (Paragraph 12)

❑ Make certain that the tenant will be able to use the office seven days a week, twenty-four hours a day. (Paragraph 12)

❑ Review the certificate of occupancy as well as the lease before agreeing to occupy the premises in accordance with the lease, the building's certificate of occupancy, and all relevant laws. (Paragraph 12)

❑ Determine who will provide and pay for utilities and services, such as air conditioning, heat, water, electricity, cleaning, window cleaning, and refuse removal. (Paragraph 13)

❑ The landlord will want the right to gain access to the premises, especially for repairs or in the event of an emergency. (Paragraph 14)

❑ Decide whether the landlord and its employees are trustworthy enough to be given keys to the premises, which has the advantage of avoiding a forced entry in the event of an emergency. (Paragraph 14)

❑ Make the landlord an additional named insured for the tenant's liability insurance policy. (Paragraph 15)

❑ Fix the amount of liability insurance that the tenant must maintain at a reasonable level. (Paragraph 15)

❑ Specify that the landlord will carry casualty and fire insurance for the building. (Paragraph 15)

❑ Indicate what limit of liability, if any, the landlord is willing to accept for interruption or harm to the tenant's business. (Paragraph 15)

❑ Agree that the lease is subordinate to any mortgage or underlying lease on the building. (Paragraph 16)

❑ If the lease requires waiver of the right to trial by jury, consider whether this is acceptable.

❑ If the lease provides for payment of the legal fees of the landlord in the event of litigation, seek to have this either stricken or changed so that the legal fees of the winning party are paid.

❑ Indicate that any rider, which is an attachment to the lease to add clauses or explain the details of certain aspects of the lease such as those relating to construction, is made part of the lease.

❑ Review the standard provisions in the introductory pages and compare them with Paragraph 18.

Filling in the Form: Form 28

In the preamble fill in the date and the names and addresses of the parties. In paragraph 1 give the information about the lease between the landlord and tenant and attach a copy of that lease as exhibit A of the sublease. In paragraph 2 give the suite number, the approximate square footage, and floor for the premises as well as the address for the building. If only part of the space is to be subleased, indicate which part. In paragraph 3 give the number of years as well as the commencement and termination dates for the lease. In paragraph 4 indicate the annual rent. In paragraph 5 specify the number of months' rent that will be given as a security deposit. In paragraph 7 specify the use that the subtenant will make of the premises. In paragraph 8 indicate whether the subtenant will be excused from compliance with any of the provisions of the lease or whether some provisions will be modified with respect to the subtenant. For paragraph 10, attach the landlord's consent as

exhibit B if that is needed. For paragraph 11, if the tenant is leaving property for the use of the subtenant, attach exhibit C detailing that property. If the lease has been recorded, perhaps with the county clerk, indicate where the recordation can be found in paragraph 14. Add any additional provisions in paragraph 15. In paragraph 16 indicate which state's laws shall govern the agreement. Both parties should then sign the sublease.

Negotiation Checklist for Form 28

❑ If the subtenant is a corporation, consider whether to insist that owners or officers of the corporation will have personal liability with respect to the sublease. (See "Other Provisions" on page 111.)

❑ If possible, do not agree to rent-free months at the start of the sublease or to paying for construction for the subtenant.

❑ Specify the location and approximate square footage of the rented premises as well as the location of the building. (Paragraphs 1 and 2)

❑ Indicate the duration of the term, including starting and ending dates. (Paragraph 3)

❑ Specify that the lease shall become month-to-month after the term expires. (Paragraph 3)

❑ Disclaim any liability if, for some reason, the tenant is unable to deliver possession of the premises at the starting date of the sublease.

❑ Do not give the subtenant a right to terminate.

❑ Do not give the subtenant an option to renew.

❑ Have a right to show the space for three or six months prior to the end of the sublease.

❑ State in the sublease that the subtenant's bankruptcy or insolvency will be grounds for termination of the lease.

❑ Specify the rent and indicate when it should be paid. (Paragraph 4)

❑ Require that rent be paid to the tenant (which allows the tenant to monitor payments), unless the tenant believes that it would be safe to allow the subtenant to pay the landlord directly. (Paragraph 4)

❏ Carefully review any proposed increases in rent during the term of the lease and make certain that the sublease will at least match such increases.

❏ Indicate the amount of the security deposit. (Paragraph 5)

❏ Specify that reductions may be made to the security deposit for sums owed the tenant before the balance is returned to the subtenant. (Paragraph 5)

❏ Require the subtenant to return the premises in broom clean condition and good order except for normal wear and tear. (Paragraph 6)

❏ Specify exactly what type of businesses the tenant may operate in the premises. (Paragraph 7)

❏ Do not allow "for no other purpose" to be stricken from the use clause.

❏ Nothing should prevent the subtenant from removing its furniture and equipment that is not affixed to the premises.

❏ Require the subtenant to occupy the premises in accordance with the sublease, the lease, the building's certificate of occupancy, and all relevant laws. (Paragraph 8)

❏ Indicate if any of lease's provisions will not be binding on the subtenant or have been modified. (Paragraph 8)

❏ State clearly that tenant is not obligated to perform the landlord's duties and that the subtenant must look to the landlord for such performance. (Paragraph 9)

❏ Agree to cooperate with the subtenant in requiring the landlord to meet its obligations, subject to a right of reimbursement from subtenant for any costs or attorney's fees the tenant must pay in the course of cooperating. (Paragraph 9)

❏ If the landlord's consent must be obtained for the sublet, attach a copy of the consent to the sublease as Exhibit B. (Paragraph 10)

❏ If tenant is going to let its property be used by the subtenant, attach an inventory of this property as Exhibit C and require that the property be returned in good condition. (Paragraph 11)

❏ Have the subtenant indemnify the tenant against any claims with respect to the premises that arise after the effective date of the sublease.

❏ Do not give the subtenant the right to sub-sublet all or part of the premises.

❏ Do not give the subtenant the right to assign the sublease.

❏ If the lease has been recorded with a government office, indicate how it can be located. (Paragraph 14)

❏ Add any additional provisions that may be necessary. (Paragraph 15)

❏ Review the standard provisions in the introductory pages and compare them with Paragraph 16.

Filling in the Form: Form 29

In the preamble, fill in the date and the names and addresses of the parties. In paragraph 1 give the information about the lease between the landlord and tenant and attach a copy of that lease as exhibit A of the lease assignment. In paragraph 3 enter an amount, even a small amount such as $10, as cash consideration. For paragraph 8, attach the landlord's consent as exhibit B if that is needed. If the lease has been recorded, perhaps with the county clerk, indicate where the recordation can be found in paragraph 9. Add any additional provisions in paragraph 10. In paragraph 11 indicate which state's laws shall govern the agreement. Both parties should then sign the assignment.

Negotiation Checklist for Form 29

❏ In obtaining the landlord's consent in a simple letter, demand that the tenant not be liable for the lease and that the landlord look only to the assignee for performance. (See "Other Provisions")

❏ It would be wise to specify the amount of security deposits or any other money held in escrow by the landlord for the tenant/assignor.

❏ Consideration should exchange hands and, if the lease has extra value because of the tenant's improvements or an upturn in the rental market, the consideration may be substantial. (Paragraph 3)

❏ Require the assignee to perform as if it were the original tenant in the lease. (Paragraph 4)

❏ Have the assignee indemnify the tenant/assignor with respect to any claims and costs that may arise from the lease after the date of the assignment. (Paragraph 5)

❏ Make the assignee's obligations run to the benefit of the landlord as well as the tenant. (Paragraph 6)

❏ Indicate that the assignor has the right to assign and the premises are not encumbered in any way.

❏ If the landlord's consent must be obtained for the assignment, attach a copy of the consent to the sublease as Exhibit B. (Paragraph 8)

❏ If the lease has been recorded with a government office, indicate how it can be located. (Paragraph 9)

❏ Add any additional provisions that may be necessary. (Paragraph 10)

❏ Review the standard provisions in the introductory pages and compare them with Paragraph 11.

Other Provisions That Can Be Used with Form 28 or 29

❏ Personal Guarantee. The tenant who sublets or assigns is bringing in another party to help carry the weight of the tenant's responsibilities pursuant to the lease. If the sublessee or assignee is a corporation, the tenant will only be able to go after the corporate assets if the other party fails to meet its obligations. The tenant may, therefore, want to have officers or owners agree to personal liability if the corporation breaches the lease. One or more principals could sign a guaranty, which would be attached to and made a part of the sublease or assignment. The guaranty that follows is for a sublease but could easily be altered to be for an assignment:

Personal Guarantee. This guarantee by the Undersigned is given to induce _____ (hereinafter referred to as the "Tenant") to enter into the sublease dated as of the _____ day of _____, 20_____, between the Tenant and _____ (hereinafter referred to as the "Sublessee"). That sublease would not be entered into by the Tenant without this guaranty, which is made part of the sublease agreement. The relationship of the Undersigned to the Sublessee is as follows _____. Undersigned fully and unconditionally guarantees to the Tenant the full payment of all rent due and other amounts payable pursuant to the Sublease. The Undersigned shall remain fully bound on this guarantee regardless of any extension, modification, waiver, release, discharge of or substitution any party with respect to the Sublease. The Undersigned waives any requirement of notice and, in the event of default, may be sued directly without any requirement that the Tenant first sue the Sublessee. In addition, the Undersigned guarantees the payment of all attorneys' fees and costs incurred in the enforcement of this guar-anty. This guaranty is unlimited as to amount or duration and may only be modified or terminated by a written instrument signed by all parties to the Sublease and Guaranty. This guaranty shall be binding and inure to the benefit of the parties hereto, their heirs, successors, assigns, and personal representatives.

❏ Assignor's Obligations. In the case of an assignment, the tenant/assignor may ask the landlord for a release from its obligations pursuant to the lease. If the landlord gives such a release, perhaps in return for some consideration or because the new tenant has excellent credentials, the tenant would want this to be included in a written instrument. This would probably be the letter of consent that the landlord would have to give to permit the assignment to take place. Shown here are options to release the tenant and also to do the opposite and affirm the tenant's continuing obligations.

Assignor's Obligations. Check the applicable provision:

❑ This Agreement relieves and discharges the Assignor from any continuing liability or obligation with respect to the Lease after the effective date of the lease assignment.

❑ This Agreement does not relieve, modify, discharge, or otherwise affect the obligations of the Assignor under the Lease and the direct and primary nature thereof.

Commercial Lease

AGREEMENT, dated the _____ day of _____, 20 _____, between
_____ (hereinafter referred to as the "Tenant"),
whose address is _____
and _____ (hereinafter referred to as the "Landlord"),
whose address is _____.

WHEREAS, the Tenant wishes to rent premises for office use;

WHEREAS, the Landlord has premises for rental for such office use;

NOW, THEREFORE, in consideration of the foregoing premises and the mutual covenants hereinafter set forth and other valuable considerations, the parties hereto agree as follows:

1. Demised premises. The premises rented hereunder are Suite # _____, comprising approximately _____ square feet, on the _____ floor of the building located at the following address _____ in the city or town of _____ in the state of _____.

2. Term. The term shall be for a period of _____ years (unless the term shall sooner cease and expire pursuant to the terms of this Lease), commencing on the ____ day of _____, 20__, and ending on the ____ day of _____, 20__. At the expiration of the term and any renewals thereof pursuant to Paragraph 3, the Lease shall become a month-to-month tenancy with all other provisions of the Lease in full force and effect.

3. Option to Renew. The Tenant shall have an option to renew the lease for a period of ____ years. Such option must be exercised in writing prior to the expiration of the term specified in Paragraph 2. During such renewal term, all other provisions of the Lease shall remain in full force and effect except for the following modifications

_____

4. Termination. The Tenant shall retain the right to terminate this Lease at the end of any month by giving the Landlord thirty (30) days written notice.

5. Rent. The annual rent shall be $_____, payable in equal monthly installments on the first day of each month to the Landlord at the Landlord's address or such other address as the Landlord may specify. The first month's rent shall be paid on signing this Lease.

6. Security Deposit. The security deposit shall be in the amount of ____ month(s) rent and shall be paid by check at the time of signing this Lease. The security deposit shall be increased at such times as the monthly rent increases, so as to maintain the security deposit at the level of ____ month(s) rent. The security deposit shall be kept in a separate interest-bearing account with interest payable to the Tenant. The security deposit, after reduction for any sums owed to the Landlord, shall be returned to the Tenant within ten days of the Tenant's vacating the premises.

7. Condition of Premises. The Tenant has inspected the premises and accepts them in "as is" condition except for the following work to be performed by the Landlord _____ _____ and completed by _____, 20___. At the termination of the Lease, The Tenant shall remove all its property and return possession of the premises broom clean and in good order and condition, normal wear and tear excepted, to the Landlord.

8. Use. The Tenant shall use and occupy the premises for _____ and for no other purpose.

9. Alterations. The Tenant shall obtain the Landlord's written approval to make any alterations that are structural or affect utility services, plumbing, or electric lines, except that the Landlord consents to the following alterations _____ _____. In addition, the Landlord consents to the installation by the Tenant of the following equipment and fixtures _____ _____ _____.

The Landlord shall own all improvements affixed to the premises, except that the Tenant shall own the following improvements _____ _____ and have the right to either remove them or sell them to a new incoming tenant. If the Landlord consents to the ownership and sale of improvements by the Tenant, such sale shall be conducted in the following way _____. If the Landlord consents to the Tenant's ownership and right to remove certain improvements, the Tenant shall repair any damages caused by such removal. Nothing contained herein shall prevent the Tenant from removing its trade fixtures, moveable office furniture and equipment, and other items not affixed to the premises.

10. Repairs. Repairs to the building and common areas shall be the responsibility of the Landlord. In addition, repairs to the premises shall be the responsibility of the Landlord except for the following _____ _____.

11. Assignment and Subletting. The Tenant shall have the right to assign this Lease to an assignee of good character and sound finances subject to obtaining the written approval of the Landlord, which approval shall not be unreasonably withheld. In addition, the Tenant shall have the right to sublet all or a portion of the premises on giving written notice to the Landlord.

12. Quiet Enjoyment. The Tenant may quietly and peaceably enjoy occupancy of the premises. The Tenant shall have access to the premises at all times and, if necessary, shall be given a key to enter the building. The Tenant shall use and occupy the premises in accordance with this Lease, the building's certificate of occupancy, and all relevant laws.

13. Utilities and Services. During the heating season the Landlord at its own expense shall supply heat to the premises at all times on business days and on Saturdays and Sundays. The Landlord ❑ shall ❑ shall not supply air conditioning for the premises at its own expense. The Tenant shall pay the electric bills for the meter for the premises, unless another arrangement exists as follows _____. _____. The Landlord shall provide and pay for water for the premises, unless another arrangement exists as follows _____ _____. The Tenant shall be responsible for and pay for having its own premises cleaned, including the securing of licensed window cleaners. ❑ The Tenant ❑ The Landlord shall be responsible to pay for the removal of refuse from the premises. If the Landlord is responsible to pay for such removal, the Tenant shall comply with the Landlord's reasonable regulations regarding the manner, time, and location of refuse pickups.

14. Access to Premises. The Landlord and its agents shall have the right, upon reasonable notice to the Tenant, to enter the premises to make repairs, improvements, or replacements. In the event of emergency, the Landlord and its agents may enter the premises without notice to the Tenant. The Tenant [] shall [] shall not provide the Landlord with keys for the premises.

15. Insurance. The Tenant agrees to carry liability insurance and name the Landlord as an additional named insured under its policy and furnish to the Landlord certificates showing liability coverage of not less than $_____ for the premises. Such company shall give the Landlord ten (10) days notice prior to cancellation of any such policy. Failure to obtain or keep in force such liability insurance shall allow the Landlord to obtain such coverage and charge the amount of premiums as additional rent payable by the Tenant. The Landlord agrees to carry casualty and fire insurance on the building, but shall not have any liability in excess of $_____ with respect to the operation of the Tenant's business.

16. Subordination. This Lease is subordinate and subject to all ground or underlying leases and any mortgages that may now or hereafter affect such leases or the building of which the premises are a part. The operation of this provision shall be automatic and not require any further consent from the Tenant. To confirm this subordination, the Tenant shall promptly execute any documentation that the Landlord may request.

17. Rider. Additional terms may be contained in a Rider attached to and made part of this Lease.

18. Miscellany. This Lease shall be binding upon the parties hereto, their heirs, successors, assigns, and personal representatives. This Agreement constitutes the entire understanding between the parties. Its terms can be modified only by an instrument in writing signed by both parties unless specified to the contrary herein. A waiver of a breach of any of the provisions of this Agreement shall not be construed as a continuing waiver of other breaches of the same or other provisions hereof. This Agreement shall be governed by the laws of the State of _____.

IN WITNESS WHEREOF, the parties hereto have signed this Agreement as of the date first set forth above.

Tenant _____
              Company Name

By _____
              Authorized Signatory, Title

Landlord _____
              Company Name

By _____
              Authorized Signatory, Title

Sublease

AGREEMENT, dated the _____ day of _____, 20 _____, between _____ (hereinafter referred to as the "Tenant"), whose address is _____ and _____ (hereinafter referred to as the "Subtenant"), whose address is _____.

WHEREAS, the Tenant wishes to sublet certain rental premises for office use;

WHEREAS, the Subtenant wishes to occupy such rental premises for such office use;

NOW, THEREFORE, in consideration of the foregoing premises and the mutual covenants hereinafter set forth and other valuable considerations, the parties hereto agree as follows:

1. **The Lease.** The premises are subject to a Lease dated as of the ____ day of _____, 20__, between _____ (referred to therein as the "Tenant") and _____ (referred to therein as the "Landlord") for the premises described as _____ and located at _____. A copy of the Lease is attached hereto as Exhibit A and made part hereof. Subtenant shall have no right to negotiate with the Landlord with respect to the Lease.

2. **Demised premises.** The premises rented hereunder are Suite # _____, comprising approximately _____ square feet, on the _____ floor of the building located at the following address _____ in the city or town of _____ in the state of _____. If only a portion of the premises subject to the Lease are to be sublet, the sublet portion is specified as follows _____.

3. **Term.** The term shall be for a period of _____ years (unless the term shall sooner cease and expire pursuant to the terms of this Sublease), commencing on the ____ day of _____, 20__, and ending on the ____ day of _____, 20__. At the expiration of the term and any renewals thereof pursuant to Paragraph 3, the Sublease shall become a month to month tenancy with all other provisions of the Sublease in full force and effect.

4. **Rent.** The annual rent shall be $_____, payable in equal monthly installments on the first day of each month to the Tenant at the Tenant's address or such other address as the Tenant may specify. The Subtenant shall also pay as additional rent any other charges that the Tenant must pay to the Landlord pursuant to the lease. Subtenant shall make payments of rent and other charges to the Tenant only and not to the Landlord. The first month's rent shall be paid to the Tenant on signing this Sublease.

5. **Security Deposit.** The security deposit shall be in the amount of ___ month(s) rent and shall be paid by check at the time of signing this Sublease. The security deposit shall be increased at such times as the monthly rent increases, so as to maintain the security deposit at the level of ___ month(s) rent. The security deposit, after reduction for any sums owed to the Tenant, shall be returned to the Subtenant within ten days of the Subtenant's vacating the premises.

6. **Condition of Premises.** Subtenant has inspected the premises and accepts them in "as is" condition. At the termination of the Sublease, Subtenant shall remove all its property and return possession of the premises broom clean and in good order and condition, normal wear and tear excepted, to the Tenant.

7. **Use.** The Subtenant shall use and occupy the premises for _____ and for no other purpose.

8. **Compliance.** Subtenant shall use and occupy the premises in accordance with this Sublease. In addition, the Subtenant shall obey the terms of the Lease (and any agreements to which the Lease, by its terms, is subject), the building's certificate of occupancy, and all relevant laws. If the Subtenant is not to be subject to certain terms of the Lease or if any terms of the Lease are to be modified for purposes of this Sublease, the specifics are as follows _____

9. **Landlord's Duties.** The Tenant is not obligated to perform the duties of the Landlord pursuant to the Lease. Any failure of the Landlord to perform its duties shall be subject to the Subtenant dealing directly with the Landlord until the Landlord fulfills its obligations. Copies of any notices sent to the Landlord by the Subtenant shall also be provided to the Tenant. The Tenant shall cooperate with Subtenant in the enforcement of Subtenant's rights against the Landlord, but any costs or fees incurred by the Tenant in the course of such cooperation shall be reimbursed by the Subtenant pursuant to Paragraph 12.

10. **Landlord's Consent.** If the Landlord's consent must be obtained for this Sublease, that consent has been obtained by the Tenant and is attached hereto as Exhibit B and made part hereof.

11. **Inventory.** An inventory of the Tenant's fixtures, furnishings, equipment, and other property to be left in the premises is attached hereto as Exhibit C and made part hereof. Subtenant agrees to maintain these items in good condition and repair and to replace or reimburse the Tenant for any of these items that are missing or damaged at the termination of the subtenancy.

12. **Indemnification.** The Subtenant shall indemnify and hold harmless the Tenant from any and all claims, suits, costs, damages, judgments, settlements, attorney's fees, court costs, or any other expenses arising with respect to the Lease subsequent to the date of this Agreement. The Tenant shall indemnify and hold harmless the Subtenant from any and all claims, suits, costs, damages, judgments, settlements, attorney's fees, court costs, or any other expenses arising with respect to the Lease prior to the date of this Agreement.

13. **Assignment and Subletting.** Subtenant shall not have the right to assign this Lease or to sublet all or a portion of the premises without the written approval of the Landlord.

14. **Recordation.** The Lease has been recorded in the office of _____ on the ___ day of _____, 20___, and is located at _____
_____.

15. **Additional provisions:**

_____
_____
_____
_____.

16. **Miscellany.** This Lease shall be binding upon the parties hereto, their heirs, successors, assigns, and personal representatives. This Agreement constitutes the entire understanding between the parties. Its terms can be

modified only by an instrument in writing signed by both parties unless specified to the contrary herein. A waiver of a breach of any of the provisions of this Agreement shall not be construed as a continuing waiver of other breaches of the same or other provisions hereof. This Agreement shall be governed by the laws of the State of

_____.

IN WITNESS WHEREOF, the parties hereto have signed this Agreement as of the date first set forth above.

Subtenant _____
Company Name

Tenant _____
Company Name

By_____
Authorized Signatory, Title

By_____
Authorized Signatory, Title

Lease Assignment

AGREEMENT, dated the _____ day of _____, 20 _____, between _____ (hereinafter referred to as the "Assignor"), whose address is _____

_____

and _____ (hereinafter referred to as the "Assignee"), whose address is _____.

WHEREAS, the Assignor wishes to assign its Lease to certain rental premises for office use;

WHEREAS, the Assignee wishes to accept the Lease assignment of such rental premises for such office use;

NOW, THEREFORE, in consideration of the foregoing premises and the mutual covenants hereinafter set forth and other valuable considerations, the parties hereto agree as follows:

1. The Lease. The Lease assigned is dated as of the _____ day of _____, 20_____, between _____ (referred to therein as the "Tenant") and _____ (referred to therein as the "Landlord") for the premises described as _____ and located at _____.

A copy of the Lease is attached hereto as Exhibit A and made part hereof.

2. Assignment. The Assignor hereby assigns to the Assignee all the Assignor's right, title, and interest in the Lease and any security deposits held by the Landlord pursuant to the Lease.

3. Consideration. The Assignee has been paid $_____ and other good and valuable consideration for entering into this Agreement.

4. Performance. As of the date of this Agreement the Assignee shall promptly perform all of the Tenant's obligations pursuant to the Lease, including but not limited to the payment of rent, and shall assume full responsibility as if the Assignee had been the Tenant who entered into the Lease.

5. Indemnification. The Assignor shall indemnify and hold harmless the Assignee from any and all claims, suits, costs, damages, judgments, settlements, attorney's fees, court costs, or any other expenses arising with respect to the Lease subsequent to the date of this Agreement. The Assignee shall indemnify and hold harmless the Assignor from any and all claims, suits, costs, damages, judgments, settlements, attorney's fees, court costs, or any other expenses arising with respect to the Lease prior to the date of this Agreement.

6. Benefited Parties. The Assignee's obligations assumed under this Agreement shall be for the benefit of the Landlord as well as the benefit of the Assignor.

7. Right to Assign. The Assignor warrants that the premises are unencumbered by any judgments, liens, executions, taxes, assessments, or other charges; that the Lease has not been modified from the version shown as Exhibit A; and that Assignor has the right to assign the Lease.

8. Landlord's Consent. If the Landlord's consent must be obtained prior to assignment of the Lease, that consent has been obtained by the Assignor and is attached hereto as Exhibit B and made part hereof.

9. **Recordation.** The Lease has been recorded in the office of _____ on the ____ day of _____, 20___, and is located at _____.

10. **Additional provisions:**

_____
_____
_____
_____

11. **Miscellany.** This Agreement shall be binding upon the parties hereto, their heirs, successors, assigns, and personal representatives. This Agreement constitutes the entire understanding between the parties. Its terms can be modified only by an instrument in writing signed by both parties unless specified to the contrary herein. A waiver of a breach of any of the provisions of this Agreement shall not be construed as a continuing waiver of other breaches of the same or other provisions hereof. This Agreement shall be governed by the laws of the State of _____.

IN WITNESS WHEREOF, the parties hereto have signed this Agreement as of the date first set forth above.

Assignee_____
       Company Name

Assignor _____
        Company Name

By_____
   Authorized Signatory, Title

By_____
   Authorized Signatory, Title

Employment Application
Employment Agreement
Restrictive Covenant
for Employment

FORM 30 FORM 31 FORM 32

Hiring new employees should invigorate and strengthen the industrial design firm. However, proper management practices have to be followed, or the results can be disappointing and, potentially, open the firm to the danger of litigation. The process of advertising a position, interviewing, and selecting a candidate should be shaped in such a way that the firm maintains clarity of purpose and fortifies its legal position.

The hiring process should avoid violating any state or federal law prohibiting discrimination, including discrimination on the basis of race, religion, age, sex, or disability. Help-wanted advertising, listings of a position with an employment agency, application forms to be filled in, questions asked during an interview, statements in the employee handbook or office manual, and office forms should all comply with these antidiscrimination laws.

The firm should designate an administrator to handle human resources. That person should develop familiarity with the legal requirements and review the overall process to protect the firm and ensure that the best candidates are hired. All employment matters—from résumés, applications, interview reports, and documentation as to employment decisions to personnel files and employee benefit information—should channel through this person.

The human resources administrator should train interviewers with respect both to legalities and the goals of the firm. Firms that lack job descriptions may not realize that this is detrimental to the firm, as well as the potential employee. Not only will the employee have a more difficult time understanding the nature of the position, but the interviewer will have a harder task of developing a checklist of desirable characteristics to look for in the interview and use as a basis of comparison among candidates. This vagueness may encompass not only the duties required in the position, but also the salary, bonuses, benefits, duration, and grounds for discharge with respect to the position. Such ambiguity is likely to lead to dissatisfaction on both sides, which is inimical to a harmonious relationship and productive work environment.

The design firm should keep in mind that employment relationships are terminable at will by either the employee or the firm, unless the firm has promised that the employment will have a definite duration. Unless the firm intends to create a different relationship, it should take steps throughout the hiring process to make clear that the employment is terminable at will. So, for example, Form 30 indicates this in the declaration signed by the applicant. The administrator should make certain that nothing in the advertisements, application, job description, interview, employee handbook, office manual, or related documents give an impression of long-term or permanent employment. Interviewers, who may be seeking to impress the applicant with the pleasant ambiance and creative culture of the firm, should not make statements such as, "Working here is a lifetime career."

The application also offers the opportunity to inform the applicant that false statements are grounds for discharge. In addition, the applicant gives permission to contact the various employers, educational institutions, and references listed in the application.

Because the employment application presents the design firm with an opportunity to support various goals in the employment process, the application is used in addition to the résumé that the employee would also be expected to make available. The application, of course, is only the beginning of the relationship between employer and employee, but it starts the relationship in a proper way that can be bolstered by the interview, the employee handbook, and related policies that ensure clarity with respect to the employee's duties and conditions of employment.

The employment agreement (Form 31) evolves from the process of application and interviews. It allows the design firm to reiterate that the employment is terminable at will. It also clarifies for both parties the terms of the employment, such as the duties, salary, benefits, and reviews. By clarifying the terms, it minimizes the likelihood of misunderstandings or disputes. It can also provide for arbitration in the event that disputes do arise.

The design firm may want to protect itself against the possibility that an employee will leave and go into competition with or harm the firm. For example, the employee might start his or her own business and try to take away clients, work for a competitor, or sell confidential information. The design firm can protect against this by the use of a restrictive covenant, which can either be part of the employment agreement or a separate agreement (Form 32). The restriction must be reasonable in terms of duration and geographic scope. It would be against public policy to allow a firm to ban an employee from ever again pursuing a career in interior design. However, a restriction that for six months or a year after leaving the firm, the employee will neither work for competitors nor start a competing firm in the same city would have a likelihood of being enforceable. Laws vary from state to state. It is wise to provide separate or additional consideration for the restrictive covenant to lessen the risk of its being struck down by a court (such as allocating a portion of wages to the restrictive covenant). This is best done at the time of first employment, since asking someone who is currently employed to sign a restrictive covenant may raise a red flag about relative bargaining positions and the validity of the consideration.

Certainly these agreements may be done as letter agreements, which have a less formal feeling and may appear more inviting for an employee to sign. For that reason, Forms 31 and 32 take the form of letters to the employee. Regardless of whether letters countersigned by the employee or more formal agreements are used, the employment agreement and the restrictive covenant should be reviewed by an attorney with expertise in employment law.

Two helpful books for developing successful programs with respect to employment are *From Hiring to Firing* and *The Complete Collection of Legal Forms for Employers*, both written by Steven Mitchell Sack (Legal Strategies Publications).

Filling in the Form: Form 30
This is filled in by the employee and is self-explanatory.

Filling in the Form: Form 31
Using the design firm's stationery, give the date and name and address of the prospective employee. In the opening paragraph, fill in the name of the design firm. In paragraph 1, give the position for which the person is being hired and the start date. In paragraph 2, indicate the duties of the position. In paragraph 3, give the annual compensation. In paragraph 4, fill in the various benefits.

In paragraph 8, indicate who would arbitrate, where arbitration would take place, and consider inserting the local small claims court limit, so small amounts can be sued for in that forum (assuming the design firm is eligible to sue in the local small claims court). In paragraph 9, fill in the state whose laws will govern the agreement. The design firm should sign two copies, the employee should countersign, and each party should keep one copy of the letter.

Filling in the Form: Form 32
Using the design firm's stationery, give the date and name and address of the prospective employee. In the opening paragraph, fill in the date of the employment agreement (which is likely to be the same date as this letter) and the name of design firm. In paragraph 1, enter a number of months. In paragraph 3, give the additional compensation that the design firm will pay. In paragraph 6, fill in the state whose

laws will govern the agreement. The design firm should sign two copies, the employee should countersign, and each party should keep one copy of the letter.

Negotiation Checklist

❏ Appoint a human resources administrator for the firm.

❏ Avoid anything that may be discriminatory in advertising the position, in the application form, in the interview, and in the employee handbook and other employment-related documents.

❏ Retain copies of all advertisements, including records of how many people responded and how many were hired.

❏ In advertisements or job descriptions, refer to the position as "full-time" or "regular," rather than using words that imply the position may be long-term or permanent.

❏ Do not suggest that the position is secure (such as "career path" or "long-term growth").

❏ Never make claims about guaranteed earnings that will not, in fact, be met.

❏ Don't require qualifications beyond what are necessary for the position, since doing so may discriminate against people with lesser qualifications who could do the job.

❏ Make certain the qualifications do not discriminate against people with disabilities who might, nonetheless, be able to perform the work.

❏ Carefully craft job descriptions to aid both applicants and interviewers.

❏ Train interviewers, and monitor their statements and questions to be certain the firm is in compliance with antidiscrimination laws.

❏ Be precise in setting forth the salary, bonuses, benefits, duration, and grounds for discharge with respect to the position.

❏ Make certain the candidates have the immigration status to work legally in the United States.

❏ Do not ask for church references while in the hiring process.

❏ Avoid asking for photographs of candidates for a position.

❏ Never allow interviewers to ask questions of women they would not ask of men.

❏ Never say to an older candidate that he or she is "overqualified."

❏ Instruct interviewers never to speak of "lifetime employment" or use similar phrases.

❏ Have the candidate give permission for the firm to contact references, prior employers, and educational institutions listed in the application.

❏ When contacting references and others listed in the application, make certain not to slander or invade the privacy of the applicant.

❏ Have the candidate acknowledge his or her understanding that the employment is terminable at will.

❏ Stress the importance of truthful responses, and have the candidate confirm his or her knowledge that false responses will be grounds for dismissal.

❏ Always get back to applicants who are not hired and give a reason for not hiring them. The reason should be based on the job description, such as, "We interviewed many candidates and hired another individual whose background and skills should make the best match for the position."

❏ Consult an attorney with expertise in the employment field before inquiring into arrests, asking for polygraph tests, requesting preemployment physicals, requiring psychological or honesty tests, investigating prior medical history, or using credit reports, since this behavior may be illegal.

❏ Create an employee handbook that sets forth all matters of concern to employees, from benefits to standards for behavior in the workplace to evaluative guidelines with respect to performance.

❏ Consider whether the firm wants to have a restrictive covenant with the employee.

❏ If a restrictive covenant is to be used, decide what behavior would ideally be preventable—working for a competitor, creating a competitive business, contacting the design firm's clients, inducing other employees to leave, or using confidential information (such as customer lists or trade secrets).

❏ Give a short duration for the covenant, such as six months or one year after termination.

❏ Make clear that additional consideration was paid to the employee for agreeing to the restrictive covenant.

❏ If there is a breach of the restrictive covenant, consult an attorney, and immediately put the employee on notice of the violation.

❏ Consider whether the firm wants an arbitration clause with the employee, since such a clause allows the firm to determine where and before which arbitrators the arbitration will take place.

❏ Make certain that any contract confirming employment specifies that the employment is terminable at will (unless the designer wishes to enter into a different arrangement). It should also clarify the duties of the employee, set forth the salary, benefits, and related information, provide for arbitration if desired, and delineate any restrictive covenant (unless that is to be in a separate document entered into at the same time).

Employment Application

Date _____

Applicant's Name _____

Address _____

Daytime telephone _____
E-mail address _____

Social Security Number
_____

Position for which you are applying
_____

How did you learn about this position?
_____

Are you 18 years of age or older?   ❏ Yes   ❏ No

If you are hired for this position, can you provide written proof that you may legally work in the United States?
❏ Yes   ❏ No

On what date are you able to commence work?
_____

Employment History

Are you currently employed?   ❏ Yes   ❏ No

Starting with your current or most recent position, give the requested information:

1. Employer _____

Address _____

Supervisor _____

Telephone number _____
E-mail address _____

Dates of employment _____
Salary _____

Description of your job title and duties
_____

Reason for leaving position

_____

May we contact this employer for a reference?   ❏ Yes   ❏ No

2.   Employer _____

Address _____

Supervisor _____

Telephone number _____
E-mail address _____

Dates of employment _____
Salary _____

Description of your job title and duties
_____

Reason for leaving position
_____

May we contact this employer for a reference?   ❏ Yes   ❏ No

3.   Employer _____

Address _____

Supervisor _____

Telephone number _____
E-mail address _____

Dates of employment _____
Salary _____

Description of your job title and duties
_____

Reason for leaving position
_____

May we contact this employer for a reference?   ❏ Yes   ❏ No

List and explain any special skills relevant for the position that you have acquired from your employment or other activities (include computer software in which you are proficient)
_____
_____
_____
_____

Educational History

Name and Address of School	Study Specialty	Number of Years Completed	Degree or Diploma
High School			
College			
Graduate School			
Other Education			

Describe any internships, other specialized training (including job-related experience in the United States military), extracurricular activities, licenses, or degrees that would be particularly helpful in performing this position

_____

_____

_____

_____

_____

To make inquiries about your work record, do we need any information about your name or your use of another?
❑ Yes   ❑ No   If you answer yes, please explain

_____

_____

References

1. Name _____
   Telephone _____
   Address _____
   E-mail address _____

How long and in what context have you known this reference?

_____

2. Name _____
   Telephone _____
   Address _____
   E-mail address _____

How long and in what context have you known this reference?

_____

3. Name _____

    Telephone _____

    Address _____

    E-mail address _____

How long and in what context have you known this reference?

_____

Applicant's Declaration

I understand that the information given in this employment application will be used in determining whether or not I will be hired for this position. I have made certain to give only true answers and understand that any falsification or willful omission will be grounds for refusal of employment or dismissal.

I understand that the employer hires on an employment-at-will basis, which employment may be terminated either by me or the employer at any time with or without cause for any reason consistent with applicable state and federal law. If I am offered the position for which I am applying, it will be employment-at-will unless a written instrument signed by an authorized executive of the employer changes this.

I know that this application is not a contract of employment. I am lawfully authorized to work in the United States and, if offered the position, will give whatever documentary proof of this as the employer may request.

I further understand that the employer may investigate and verify all information I have given in this application, on related document (including but not limited to my resume), and in interviews. I authorize all individuals, educational institutions, and companies named in this application to provide any information the employer may request about me, and I release them from any liability for damages for providing such information.

Applicant's signature _____   Date _____

Employment Agreement

[Designer's Letterhead]

Date _____

Mr./Ms. [New Employee]
_____ [address]
_____

Dear :

We are pleased that you will be joining us at _____ (hereinafter referred to as the "Company"). This letter is to set forth the terms and conditions of your employment.

1. Your employment as _____ shall commence on _____, 20___.

2. Your duties shall consist of the following
_____
_____
_____

You may also perform additional duties incidental to the job description. You shall faithfully perform all duties to the best of your ability. This is a full-time position and you shall devote your full and undivided time and best efforts to the business of the Company.

3. You will be paid annual compensation of $_____ pursuant to the Company's regular handling of payroll.

4. You will have the following benefits:

a) Sick days _____

b) Personal days _____

c) Vacation _____

d) Bonus _____

e) Health Insurance _____

f) Retirement Benefits _____

g) Other _____

5. You will familiarize yourself with the Company's rules and regulations for employees and follow them during your employment.

6. This employment is terminable at will at any time by you or the Company.

7.   You acknowledge that a pre-condition to this employment is that you negotiate and sign a restrictive covenant prior to the commencement date set forth in Paragraph 1.

8.   Arbitration. All disputes arising under this Agreement shall be submitted to binding arbitration before _____ in the following location _____ and settled in accordance with the rules of the American Arbitration Association. Judgment upon the arbitration award may be entered in any court having jurisdiction thereof. Disputes in which the amount at issue is less than $_____ shall not be subject to this arbitration provision.

9.   Miscellany. This agreement shall be binding on both us and you, as well as heirs, successors, assigns, and personal representatives. This agreement constitutes the entire understanding. Its terms can be modified only by an instrument in writing signed by both parties. Notices shall be sent by certified mail or traceable overnight delivery to you or the Company at our present addresses, and notification of any change of address shall be given prior to that change of address taking effect. A waiver of a breach of any of the provisions of this agreement shall not be construed as a continuing waiver of other breaches of the same or other provisions hereof. This agreement shall be governed by the laws of the State of _____.

If this letter accurately sets forth our understanding, please sign beneath the words "Agreed to" and return one copy to us for our files.

Sincerely yours,   Agreed to:

_____   _____
          Company Name   Employee

By _____
          Name, Title

Restrictive Covenant for Employment

[Designer's Letterhead]

Date _____

Mr./Ms. [New Employee]
_____ [address]
_____

Dear :

By a separate letter dated _____, 20__, we have set forth the terms for your employment with _____ (hereinafter referred to as the "Company").

This letter is to deal with your role regarding certain sensitive aspects of the Company's business. Our policy has always been to encourage our employees, when qualified, to deal with our clients and, when appropriate, contact our clients directly. In addition, during your employment with the Company, you may be given knowledge of proprietary information that the Company wishes to keep confidential.

To protect the Company and compensate you, we agree as follows:

1.   You will not directly or indirectly compete with the business of the Company during the term of your employment and for a period of ____ months following the termination of your employment, regardless of who initiated the termination, unless you obtain the Company's prior written consent. This means that you will not be employed by, own, manage, or consult with a business that is either similar to or competes with the business of the Company. This restriction shall be limited to the geographic areas in which the Company usually conducts its business, except that it shall apply to the Company's clients regardless of their location.

2.   In addition, you will not during the term of your employment or thereafter directly or indirectly disclose or use any confidential information of the Company except in the pursuit of your employment and in the best interest of the Company. Confidential information includes but is not limited to client lists, client files, trade secrets, financial data, sales or marketing data, plans, designs, and the like relating to the current or future business of the Company. All confidential information is the sole property of the Company. This provision shall not apply to information voluntarily disclosed to the public without restrictions or which has lawfully entered the public domain.

3.   As consideration for your agreement to this restrictive covenant, the Company will compensate you as follows

_____

_____

4.   You acknowledge that, in the event of your breach of this restrictive covenant, money damages would not adequately compensate the Company. You therefore agree that, in addition to all other legal and equitable remedies available to the Company, the Company shall have the right to receive injunctive relief in the event of any breach hereunder.

5.   The terms of this restrictive covenant shall survive the termination of your employment, regardless of the reason or causes, if any, for the termination, or whether the termination might constitute a breach of the agreement of employment.

6.   Miscellany. This agreement shall be binding on both us and you, as well as heirs, successors, assigns, and personal representatives. This agreement constitutes the entire understanding. Its terms can be modified only by an instrument in writing signed by both parties. Notices shall be sent by certified mail or traceable overnight delivery to you or the Company at our present addresses, and notification of any change of address shall be given prior to that change of address taking effect. A waiver of a breach of any of the provisions of this agreement shall not be construed as a continuing waiver of other breaches of the same or other provisions hereof. This agreement shall be governed by the laws of the State of _____.

If this letter accurately sets forth our understanding, please sign beneath the words "Agreed to" and return one copy to us for our files.

Sincerely yours,   Agreed to:

_____   _____
           Company Name   Employee

By _____
           Name, Title

Project Employee Contract

An industrial design firm may want to hire someone who falls into an intermediate position between an independent contractor (discussed with respect to Form 34) and a permanent employee. If the person is to work on a project for a period of months, it is quite likely that he or she meets the legal definition of an employee. While the person may meet the IRS tests for who is an employee, the design firm may prefer not to give the person the full benefits of the typical employment contract.

The first consideration in such a situation is whether the person could be hired as an independent contractor. Since IRS reclassification from independent contractor to employee can have harsh consequences for the design firm, including payment of back employment taxes, penalties, and jeopardy to qualified pension plans, the IRS has promulgated guidelines for who is an employee.

Basically, an employee is someone who is under the control and direction of the employer to accomplish work. The employee is not only told what to do, but how to do it. On the other hand, an independent contractor is controlled or directed only as to the final result, not as to the means and method to accomplish that result. Some twenty factors enumerated by the IRS dictate the conclusion as to whether someone is an employee or an independent contractor, and no single factor is controlling. Factors suggesting someone is an independent contractor would include that the person supplies his or her own equipment and facilities; that the person works for more than one party (and perhaps employs others at the same time); that the person can choose the location to perform the work; that the person is not supervised during the assignment; that the person receives a fee or commission rather than an hourly or weekly wage; that the person can

make a loss or a profit; and that the person can be forced to terminate the job for poor performance, but cannot be dismissed like an employee. The designer should consult his or her accountant or attorney to resolve any doubts about someone's status.

Assuming that these criteria suggest that a person to be hired for a project is an employee, the design firm may choose to designate him or her as a project employee. Project employees are usually hired for a minimum of four months. They may be transferred from one assignment to another. Project employees are usually eligible for most benefits offered other regular employees, such as medical/dental benefits, life insurance, long-term disability, vacation, and so on. However, project employees would not be eligible for leaves of absence and severance pay. If a project employee moves to regular status, the length of service will usually be considered to have begun on the original hire date as a project employee.

Filling in the Form

This contract would normally take the form of a letter written on the company's letterhead. Fill in the name of the project employee in the salutation. Then, specify the type of work that the project employee will be doing—i.e., industrial designer, assistant, account executive, and so on. Indicate the start date for work and the anticipated project termination date. In the second paragraph, give the salary on an annualized basis, as well as on a biweekly basis. In the third paragraph, state benefits for which the project employee will not be eligible, such as leaves of absence, tuition reimbursement, and severance pay. Insert the company name in the last paragraph and again after "Sincerely." Both parties should sign the letter.

Project Employee Contract

[Designer's Letterhead]

Dear :

This will confirm that you have accepted a Project Job as _____ with our company. This assignment will begin on _____, and has an expected project termination date on or before _____.

As we agreed, based on an annual salary of $_____, you will be paid $_____ on biweekly basis, less applicable taxes and insurance. During the term of the assignment, this employment will be terminable at will either by you as Project Employee or by us as Project Employer.

As a Project Employee, you are eligible for our company's benefits except for

_____

_____

_____

I cannot guarantee your employment beyond this assignment. The project employee status allows you to consider finding another job within the company. An added benefit is that if your Project Employment status changes to that of regular employee, your original Project hire date will become your start date of continuous employment.

I am very pleased to welcome you to_____ and look forward to working with you. Please let me know if you have any questions or concerns.

Sincerely yours,   Agreed to:

_____   _____
          Company Name   Project Employee

By _____   Date _____
          Project Manager

Contract with an Independent Contractor

Form 34 can be used by the designer, or by the designer's client, to contract with an independent contractor. The advantage of having the client contract directly is that the designer is not in an intermediary position, with its attendant risks. However, the designer should advise the client to consult with the client's attorney for contract forms and advice when contracting with an independent contractor, since the designer should never give or appear to give legal advice. Also, while the designer may suggest contractors to the client, it is wise to have the client make the final selection of the contractor. If the designer likes to work with particular contractors and insists the client hire among them, the designer should be prepared to be blamed by the client if anything goes wrong—even if the client has contracted directly with the contractor and the designer has no legal liability.

Independent contractors run their own businesses and hire out on a job-by-job basis. They are not employees, which saves the designer in terms of employee benefits, payroll taxes, and paperwork. By not being an employee, the independent contractor does not have to have taxes withheld and is able to deduct all business expenses directly against income.

A contract with an independent contractor serves two purposes. First, it shows the intention of the parties to have the services performed by an independent contractor. Second, it shows the terms on which the parties will do business.

As to the first purpose of the contract—showing the intention to hire an independent contractor, not an employee—the contract can be helpful if the Internal Revenue Service (IRS) decides to argue that the independent contractor was an employee. The tax law automatically classifies as independent contractors physicians, lawyers, general building contractors, and others who follow an independent trade, business, or profession, in which they

offer their services to the public on a regular basis. However, many people do not fall clearly into this group. IRS guidelines as to the employee–independent contractor distinction are discussed in relation to Form 33.

The second purpose of the contract is to specify the terms agreed to between the parties. What services will the contractor provide, and when will the services be performed? On what basis and when will payment be made by the designer? If a project fee is to be paid, will it be paid in installments so that the payment of fees is closely matched to the progression of work? Will there be an advance, perhaps to help defray expenses? Will the designer demand a provision for retainage, so that 10 to 15 percent of all payments are withheld until satisfactory completion of the work, and will the contractor agree to this?

There should be agreement about a schedule, which would include start and completion dates. Whether time is of the essence, which would require that time deadlines be strictly met, should also be indicated in the contract. The designer should consult with his or her insurance agent with respect to the insurance the designer should carry when dealing with independent contractors. Certainly, the designer should make sure there is adequate coverage for property damage and liability arising from lawsuits for injuries. The contractor should definitely have its own liability policy, as well as workers' compensation and state disability coverage. The designer may require the independent contractor to maintain a certain level of insurance in force and include the designer as an additional named beneficiary of that policy.

Independent contractors can perform large jobs or render a day's services. Form 34 is designed to help designers deal with small independent contractors who are performing a limited amount of work. The negotiation checklist is also directed toward this situation.

However, some further discussion is necessary to cover the issues arising when the designer has a larger project to complete.

If the designer were dealing with a substantial renovation or other construction, the contract would have to be more complex. First, it is always wise to have references for a contractor who is new to the designer. Keep deposits small, since it can be hard to get a deposit back if the contractor does not perform. There should also be clarity as to the quality and perhaps even the brands of any materials to be used. The contractor can be asked to post a surety bond, which is a bond to guarantee full performance. However, many small contractors may have difficulty obtaining such a bond, since the insurance company may require the posting of collateral. In any event, the designer might explore with his or her own insurance agent the feasibility of demanding this from the contractor. If the design is to be incorporated into a building, the contractor's failure to pay subcontractors or suppliers of material can result in a lien against the property where the work has been done. A lien is like a mortgage on a building; it must be satisfied or removed before the property can be sold. A surety bond would avoid problems with liens. Another approach is to require the contractor and any subcontractors to provide the designer with waiver of liens when the work is finished.

A contractor should be required to give a bid. That bid will be the basis for the terms of the contract. The contractor may want to wait until after completing the work to determine a fee. Obviously, this is unacceptable. The contractor may want to charge a fee for labor, but charge cost plus a markup for materials. This is probably also unacceptable, since the designer has a budget and needs to know that budget can be met. Another variation is for the contractor to allow for a 10 percent variation in the bid or the costs of materials based on what actually happens. This should be carefully evaluated by the designer, but is less desirable than a firm fee. The fee and the job description should only be modified by a written amendment to the contract. If this isn't required, disputes are likely to result.

The designer should, if possible, require the contractor to warrant a number of facts, such as the contractor being licensed, if necessary, the materials being new and of good quality, the contractor being responsible for any damages arising from its work, and any construction being guaranteed for some period of time. The contractor would agree to protect the designer (by paying losses, damages, and any attorney's fees) in the event any of these warranties were breached. If permits are necessary to do the work, the contract should specify who will obtain the permit.

Keep in mind that Form 34 is designed for projects of limited scale, not the hiring of contractors for major undertakings that would usually require additional specifications and legal safeguards.

Filling in the Form

In the preamble, fill in the date and the names and addresses of the parties. In paragraph 1, show in detail what services are to be performed. Attach to the contract another sheet of description or a list of procedures, phases, specifications, diagrams, or plans, if needed. If the designer is to obtain or pay for permits, fill in that information. Indicate if the contractor is to obtain a payment bond or a performance bond. In paragraph 2, give a schedule. In paragraph 3, deal with the fee and expenses. In paragraph 4, specify a time for payment.

In paragraph 5, indicate how cancellations will be handled. In paragraph 6[E], fill in any special criteria that the contractor should warrant as true. In paragraph 7, fill in the information with respect to insurance the contractor must carry. In paragraph 10, specify who will arbitrate, where the arbitration will take place, and, if local small claims court would be better than arbitration, give amounts under the small claims court dollar limit as an exclusion from arbitration. In paragraph 11, give the state whose laws will govern the contract. Have both parties sign the contract.

Negotiation Checklist

❑ Carefully detail the services to be performed. If necessary, attach an additional sheet of description, a list of procedures, phases, specifications, diagrams, or plans. (Paragraph 1)

❑ Require the contractor to obtain and pay for any needed permits. (Paragraph 1)

❑ Indicate that the contractor has taken into account when setting fees any special circumstances regarding access, location, existing conditions, and the like. (Paragraph 1)

❑ Require that when the project is completed, the contractor and any subcontractors will provide the designer with waivers of liens. (Paragraph 1)

❑ Especially for commercial projects, require the contractor to obtain a performance bond and a payment bond. (Paragraph 1)

❑ Set the standard the contractor must meet, such as "work of the highest quality." (Paragraph 1)

❑ Set the standard by which the work will be judged, such as that the work must be satisfactory to the designer. (Paragraph 1)

❑ Determine whether a certain number of workers should always have to be present for the project.

❑ Clarify that the contractor must supervise any workers. (Paragraph 1)

❑ Consider having the contractor give a general release to the designer after full payment has been made, so no question can arise about money being owed to the contractor.

❑ Give a schedule for performance. (Paragraph 2)

❑ With respect to the schedule, consider making time of the essence, so that lateness will be a breach of the contract.

❑ State the method for computing the fee. (Paragraph 3)

❑ If a project fee is used, pay in installments, so the contractor does not get paid ahead of the work. (Paragraph 3)

❑ If the contractor is to bill for expenses, limit which expenses may be charged. (Paragraph 3)

❑ Require full documentation of expenses in the form of receipts and invoices. (Paragraph 3)

❑ Place a maximum amount on the expenses that can be incurred. (Paragraph 3)

❑ Pay an advance against expenses, if the amount of the expenses is too much for the contractor to wait to be reimbursed upon completion of the entire job. (Paragraph 3)

❑ State a time for payment. (Paragraph 4)

❑ Provide that notice will be given before termination for cause, so the delay or other problem can be cured. (Paragraph 5)

❑ Indicate whether the bankruptcy or insolvency of the contractor should terminate the contract. (Paragraph 5)

❑ Decide whether the death of the designer or contractor should terminate the agreement.

❑ Consider setting a reasonable amount for "liquidated damages," such as a daily dollar amount the contractor must pay the designer if the completion deadline isn't met. (Paragraph 5)

❑ Consider offering a bonus for early completion, if there is a damages provision for late completion.

❑ Even if liquidated damages are paid, reserve the right to sue for damages. After termination, for example, the designer may have to hire a different contractor to finish the job and may end up paying the new contractor far more than the liquidated damages received from the old contractor. (Paragraph 5)

❑ If expenses are billed for, consider whether any markup should be allowed.

❑ Require warranties that the contractor is legally able to perform the contract (including being licensed, if that is necessary), that all services will be done in a professional manner, that any subcontractor or employee hired by the contractor will be professional, that the contractor will pay all taxes for the contractor and his or her employees, and any other criteria for the proper performance of the services. (Paragraph 6)

❏ Review insurance coverage with the designer's insurance agent.

❏ Specify what insurance coverage the contractor must have. (Paragraph 7)

❏ Require that the contract cover the designer as an additional named insured. (Paragraph 8)

❏ Make certain the contractor actually obtains the insurance coverage and follows through in naming the designer as an additional named insured.

❏ State that the parties are independent contractors and not employer-employee. (Paragraph 8)

❏ Do not allow assignment of rights or obligations under the contract. (Paragraph 9)

❏ The designer should check with his or her attorney as to whether arbitration is better than suing in the local courts, and whether small claims court might be better than arbitration. (Paragraph 10)

❏ If the designer has not appointed a project representative and the contractor requests this, comply with the request.

❏ Require that changes to the contract be written and signed by both parties, including work change orders.(Paragraph 11)

❏ If a work change order affects the fee, as will frequently be the case, the work change order should detail the impact of the fee. (Paragraph 11)

❏ Compare paragraph 11 with the standard provisions in the introduction.

❏ If a contractor creates a work protected by copyright, trademark, or patent law, make certain that any needed rights of usage or ownership are obtained.

Contract with an Independent Contractor

AGREEMENT, dated the _____ day of _____, 20 _____, between _____ (hereinafter referred to as the "Designer"), whose address is _____ and _____ (hereinafter referred to as the "Contractor"), whose address is _____.

The Parties hereto agree as follows:

1. **Services to be Rendered.** The Contractor agrees to perform the following services for the Designer

_____
_____
_____
_____

The Contractor has reviewed special aspects of the project, such as location, existing features, access, and the like, and has incorporated all such considerations into the fee specified in Paragraph 3.

If any permits are necessary to perform these services, the Contractor shall obtain these permits at its expenses unless provided to the contrary here

_____

The Contractor shall obtain payment and performance bonds as follows

_____
_____
_____

The Contractor shall provide work of the highest quality, which shall be completed to the satisfaction of the Designer. The Contractor shall be responsible for supervision of the work in progress. Upon completion, the Contractor shall provide the Designer with waivers of liens both for the Contractor and any subcontractors used by the Contractor.

If needed, a list of procedures, phases, specifications, diagrams, or plans for the services shall be attached to and made part of this Agreement.

2. **Schedule.** The Contractor shall complete the services pursuant to the following schedule

_____
_____

Time shall be of the essence for performance under this contract.

3. **Fee and Expenses.** The Designer shall pay the Contractor as follows
❏ Project rate $_____, in installments as follows

_____

❏ Day rate $_____ /day
❏ Hourly rate $_____ /hour
❏ Other _____

The Designer shall reimburse the Contractor only for the expenses listed here

_____
_____
_____

Expenses shall not exceed $_____. The Contractor shall provide full documentation for any expenses to be reimbursed, including receipts and invoices. An advance of $_____ against expenses shall be paid to the Contractor and recouped when Payment is made pursuant to Paragraph 4.

4. Payment. Payment shall be made: ❏ at the end of each day ❏ upon completion of the project ❏ within thirty (30) days of Designer's receipt of Contractor's invoice
❏ Other _____

5. Termination. This Agreement may be terminated at any time for cause by either party notifying the other party in writing of that party's breach of the Agreement and giving five (5) business days for a cure, after which the notifying party may terminate if there has been no cure of the breach. Causes for termination shall include but not be limited to failure to perform any duty pursuant to this Agreement in a timely manner, postponements of the Project for more than ____ business days in total, and the bankruptcy or insolvency of a party hereto. While reserving all other rights under this Agreement, including the right to sue for damages, in the event that the Project is terminated, the Designer shall have the right to be paid $____ for each day that the Contractor has fallen behind the schedule.

6. Warranties. The Contractor warrants as follows.

(A) The Contractor is fully able to enter into and perform its obligations pursuant to this Agreement, including but not limited to having any licenses that may be necessary to perform the services required hereunder.

(B) All services shall be performed in a professional manner.

(C) If the employees or subcontractors are to be hired by Contractor they shall be competent professionals.

(D) The Contractor shall pay all necessary local, state, or federal taxes, including but not limited to withholding taxes, workers' compensation, F.I.C.A., and unemployment taxes for Contractor and its employees.

(E) Any other criteria for performance are as follows

_____
_____
_____
_____

7. Insurance. The Contractor agrees to carry liability insurance and name the Designer as an additional named insured under its policy and furnish to the Designer certificates showing liability coverage of not less than $_____ applicable to the services to be performed hereunder. The insurer must be a company recognized to do business in the State of _____. Such company shall be required to give the Designer ten (10) days notice prior to cancellation of any such policy. Failure to obtain or keep in force such liability insurance shall allow the Designer to obtain such coverage and charge the amount of premiums as an offset against fees payable hereunder to the Contractor. In addition, the Contractor shall maintain in force the following other types of insurance

_____
_____

8. Relationship of Parties. Both parties agree that the Contractor is an independent contractor. This Agreement is not an employment agreement, nor does it constitute a joint venture or partnership between the Designer and Contractor. Nothing contained herein shall be construed to be inconsistent with this independent contractor relationship.

9. Assignment. This Agreement may not be assigned by either party without the written consent of the other party hereto.

10. Arbitration. All disputes shall be submitted to binding arbitration before _____ in the following location _____ and settled in accordance with the rules of the American Arbitration Association. Judgment upon the arbitration award may be entered in any court having jurisdiction thereof. Disputes in which the amount at issue is less than $_____ shall not be subject to this arbitration provision.

11. Miscellany. This Agreement constitutes the entire agreement between the parties. Its terms can be modified only by an instrument in writing signed by both parties, including but not limited to work change orders. This Agreement shall be binding on the parties, their heirs, successors, assigns, and personal representatives. A waiver of a breach of any of the provisions of this Agreement shall not be construed as a continuing waiver of other breaches of the same or other provisions hereof. This Agreement shall be governed by the laws of the State of _____.

IN WITNESS WHEREOF, the parties hereto have signed this as of the date first set forth above.

Designer _____
           Company Name

Contractor _____

By _____
    Authorized Signatory

Industrial Designer–Client Agreement
Industrial Design Contact Summary Sheet

It is a necessity for business success that contracts are entered into for performing a wide variety of professional services. This includes industrial designers who are often called upon to provide assistance in the design and development process for numerous products. There is an almost universal understanding that industrial designers add value to the competitive position of products as well as the overall customer experience.

Regardless of the task, the industrial designer should put in writing an agreement to spell out clearly the functional obligations, compensation, timelines, and other terms of performance. Although many oral agreements can be valid, a written contract is always recommended. There are some instances where the agreement must be in writing to be enforceable, such as a contract that cannot be completed within one year.

The written industrial designer–client agreement can be done in many forms; from a simple letter that is signed by the designer and the client, to a complex, lengthy, formal document that can include many attachments. Keep in mind that each industrial design project will likely have some unique features and issues, and the industrial designer–client agreement should be tailored to address these matters. However, the use of standardized and modifiable agreement forms, such as Form 31 of this book, can be extremely useful to the industrial designer in putting effective written contracts in place. The CD-ROM provided with the book makes customization of the agreement simple and easy.

After identifying the parties and reciting the consideration, Form 35 starts with a description of the project. This description includes a summary of the industrial design project as well as a detailed description that provides the scope and the specific project phases. Many industrial design projects can take a product from inception to production, and therefore the industrial designer–client agreement may require particular clauses that spell out the steps in each phase. These phases may include design research, analysis, concepts, drawings, models and prototypes. Some projects may also involve molds, manufactured parts, assembled products, packaging and additional services. Industrial design projects may require budget estimates and effective contract documentation and administration.

Good communication between the industrial designer and client is very important. The industrial designer–client agreement can provide for meetings with clients to discuss the project's objectives and status. These can be scheduled at a regular frequency or as needed throughout the design project.

Scheduling can be a very difficult part of the industrial design agreement. There may be many causes for delay that are beyond the control of the industrial designer. For example, materials and supplies may come in late, contractors and others may fail to meet their deadlines, and delays could result from natural disasters and civil unrest. The industrial designer should make sure that these factors are considered when putting together a schedule in the contract. Try to never guarantee a completion date, but instead the designer could agree to be diligent in moving the project promptly to completion. The agreement should protect the industrial designer against the failures of other parties who are involved in the project. The quality and supervision of product or manufacturing should be the responsibility of the suppliers or contractors and not that of the industrial designer. If the designer contracts with third parties, he or she can assign to the client whatever rights the designer may have against the third parties.

It is important to provide how purchases of merchandise, materials, production, and other services will be handled under the agreement. The parties need to decide if the client will arrange for purchases directly, or if the designer will make these purchases as agent for the client, or if they will be handled through other arrangements. It is always preferable for the client to make all required payments directly to the suppliers or contractors. If the designer is acting as agent for the client, the client will likely want to approve all purchases by signing a written authorization.

Although the focus has been on the industrial designer's duties and responsibilities, the client will also have responsibilities under the agreement. The client's duties may include, for example, arranging purchases as mentioned earlier; cooperating with the designer; and providing project specifications, feedback, and other information. Be sure to spell out the client's duties in the agreement.

How the industrial designer will be paid for services is a significant component of the agreement. Decide if the designer is to be paid at an hourly rate, a per diem rate, a flat fee, or other method. Where there are large upfront costs for the designer, designers typically ask for part of the remuneration as a nonrefundable fee due upon the signing of the agreement. The designer may also ask for a monthly retainer to generate a steady flow of income. If the project continues beyond the anticipated timeline, the designer might seek a fee for each additional month until completion. Whatever the circumstances, the designer and client should negotiate a method of remuneration that best suits the project.

To protect against endless changes to the design project, the designer should include in the agreement a provision that governs revisions. The designer typically agrees to make a reasonable amount of revisions requested by the client without additional charge. However, if the revisions are requested after the client has given approval, an additional fee should be charged to compensate the designer for any added time and costs. In general, if the services requested by the client exceed the scope of the original project, there should be a clause in the agreement for increasing the designer's fee. In

addition, the designer should be reimbursed for all expenses connected to the project, such as messengers, long-distance telephone calls, overnight deliveries, and travel expenses.

The manner of payment is another significant term that is included in the industrial designer–client agreement. The agreement provides that the client will pay the designer within a specified time (typically thirty to sixty days) after receipt of the designer's billings. Interest is charged on any overdue payments. If the designer has to take legal action to enforce the contract and is the winning party, the client will reimburse the designer for attorneys and other costs.

Careful consideration must be given to the termination of the agreement and what happens thereafter. The agreement provides that either party may terminate the agreement for cause after the defaulting party has failed to correct the problems within a specified time. To prevent inadvertent termination of the agreement while the project is still ongoing, the agreement renews automatically unless the parties give a thirty-day prior notice of termination. To protect the designer for his or her work performed before termination, the agreement states that the designer will be paid for the work done up to termination, as well as approved purchases of materials and production, and expenses.

The industrial designer and the client must come to a clear understanding regarding ownership of the designs created by the designer pursuant to the agreement, including intellectual property and other proprietary rights. The preferred arrangement for the designer is where the designer retains ownership of the designs, including any drawings, renderings, sketches, samples, or other materials prepared by the designer during the course of the project. This arrangement gives the designer better control over the design and its use because the designer would own any copyrights, trademarks, patents, or other proprietary rights existing in the design. The client cannot use the designs outside of the particular project without obtaining the designer's permission and paying compensation.

In many cases, particularly if large design fees are paid, the client will want to own some or all of the designs. The final outcome is a negotiated one and will depend on the relative leverage of the designer and the client. A reasonable compromise is to give the client ownership of any design in its final form that is selected and approved for use in the project by the client. However, all other designs, concepts, and ideas prepared by the designer in the course of the project would be exclusively owned by the designer. Other suitable arrangements surrounding ownership of the designs can also be worked out and put into the written agreement.

The client may choose to retain other designers and professionals on the project, such as artists, draftsmen, or engineers. In such an event the client should be responsible for supervising and paying these consultants. The designer will cooperate with these professionals but will not be responsible for their work product.

The designer may want to publicize the design work done on the project for business development purposes. The agreement should give the designer the right to document the project at his or her expense for publicity purposes. If the client documents and publicizes the project, the designer should be given credit for his or her work on the project.

The business relationship between the designer and the client should be clearly defined by the agreement. Typically, the designer works as an independent contractor and not as an employee of the client.

One commonly overlooked aspect of designer and other services contracts is the ability of the parties to assign or transfer their rights. Generally, rights under a contract are assignable unless the contract provides otherwise. However, some contracts may not be automatically assignable if they involve personal services, such as design work. The sample designer–client agreement provides that the agreement is not assignable without the written consent of the other party. But the designer can assign rights to payments due under the agreement because this should have no impact on the design services.

In many cases it is advisable to include in the industrial designer–client agreement a provision for binding arbitration or nonbinding mediation of disputes. Lawsuits can be lengthy and costly, while alternative dispute resolution such as arbitration is usually quicker and less expensive. Organizations like the American Arbitration Association have a history of success in resolving contract disputes outside of the litigation process. However, arbitration may not be appropriate in all circumstances, such as where the rights and obligations of third parties are involved or where small dollar amounts can be handled in small claims court. The designer may want to seek the advice of an attorney to determine if the agreement should contain a binding arbitration clause.

The designer should include in the agreement any additional terms and conditions that are believed to be necessary or desirable. The agreement should reflect the entire understanding of the parties, and the agreement should supercede any oral and extraneous signed statements and comments. Modifications to the agreement should only be made in writing and signed by both parties. The agreement needs to be binding on the client and designer, as well as other parties who may stand in their place, such as heirs, successors, assigns, and personal representatives. Choose the state where the agreement is to be interpreted and governed. Ideally, the governing state should be where the designer is located or most familiar.

The agreement must be properly executed by the industrial designer and the client. Only authorized signatories for both parties should sign the agreement to be certain that it is binding and legally enforceable. The designer needs to make sure that he or she is dealing with the right person at the client's business, such as the proprietor, chief executive, president, or other high-level officer. Where the designer has doubts, he or she should ask the client for proof of authorization.

After the agreement has been signed, a contract summary should be prepared. This summary provides an easy reference for the designer and others working with him to the important terms and provisions of the contract.

Filling in the Form: Form 35

To complete Form 35, start by filling in the date and the names and addresses of the client and designer. Check the appropriate box in paragraph 1 to indicate whether the design plan is to be described in the body of the paragraph or in an attached schedule A. If a schedule A is attached, it should contain a complete description of the design project. Otherwise, fill in the project description in paragraph 1, including the scope of the designer's work and the various project phases where applicable. Also check the box indicated if the designer will prepare an estimated budget. Add the frequency (e.g., weekly, monthly, etc.) that the designer and client will meet to discuss the design project. If there are additional services to be performed by the designer, include these at the end of paragraph 1.

In paragraph 2, check the appropriate box to indicate that the designer's work is to start on a specified date, or after the client has provided the identified information to the designer. Also check the applicable boxes to show the date for completion of the project and the schedule to which the designer will conform.

Complete paragraph 3 by checking the relevant boxes to indicate how purchases will be handled. Indicate whether the client will arrange for and pay for purchases, or whether the designer will act as agent to arrange the purchases, or whether other arrangements to be filled in will be followed. Also check the box to show that the designer shall or shall not prepare purchase orders for these purchases.

The client will designate a person, if necessary, to act as liaison with the designer in paragraph 5. Additionally, fill in those deliveries that the designer is responsible for receiving, inspecting and storing.

In paragraph 6 select the manner of payments to the designer. If payment is hourly or per diem, fill in the rate and indicate if the client will pay a nonrefundable, up-front design fee. Put in the amount of any monthly retainer that will be paid to the designer. If a flat fee plus a percentage of costs will be paid to the designer, fill in the percent markup

and list those items which will not be included in the markup calculation. If a flat fee is the manner of payment, include the total amount along with any upfront fee and the amount and schedule of any installment payments. Include the amount of any additional payments if the project continues beyond a specified date. Likewise in paragraph 7, include any additional fees due the designer if there are revisions after the client has already approved the design.

To handle expenses under paragraph 8, enter the percent markup on expenses that will be paid to the designer. Where nonlocal travel is involved, fill in the billing procedures for travel expenses. If the designer will be paid a nonrefundable advance against expenses, fill in the amount of the advance. For paragraph 9, insert the number of days for payment of the designer's bills after receipt thereof. Include the monthly interest charge for overdue payments.

In paragraph 10, enter the date that the agreement will expire and the additional periods for which the agreement will automatically renew unless a timely notice of termination is given. Furthermore, indicate the number of total days of postponement of the project that will be a cause for termination of the agreement.

Indicate who will have ownership of the designs by checking the appropriate box in paragraph 11. In paragraph 13, include any restrictions on the designer's use of the client's name and location for publicity purposes. In paragraph 16, fill in the name(s) of the person(s) or organization(s) which will arbitrate disputes under the agreement, as well as the location of the arbitration . Fill in the value amount of disputes which will not be subject to arbitration.

Paragraph 17 is where any additional terms and conditions are included in the contract. Complete paragraph 18 by entering the state where the agreement will be governed.

At the end of the agreement, fill in the names of the designer and the client, including their respective company or business names. Include the titles or the authorized capacity of the individuals who will sign the agreement.

Negotiation Checklist

❏ The designer should check the client's credit to make sure the client has a good history of paying bills and is able to pay for the designer's services.

❏ Clearly specify the description of the design project and the scope of the work that the designer will perform.

❏ For a complex industrial design project, decide the number of phases into which the project should be subdivided. Clarify the nature of the work and activities that is to be completed for each phase and the appropriate timelines.

❏ Decide whether the designer will prepare an estimated budget and whether the costs for preparing the budget will be included in the price for the designer's work on the project. Make sure that the client understands and acknowledges that the estimated budget is not a price guarantee, but is subject to change.

❏ Set up a timetable for periodic meetings between the designer and client that is reasonable in view of the designer's schedule.

❏ Make sure that any additional work to be performed by the designer is itemized in the agreement, and that the designer will receive additional compensation for this work.

❏ Be certain that suppliers and contractors are responsible for the quality and supervision of their work. Limit the amount of supervision that the industrial designer will provide.

❏ Always set a schedule for completion of the project that is within the control of the designer. Make certain that the designer is not held responsible for delays caused by the client, suppliers, or acts of God, and that the designer's time to perform is extended by any such delays.

❏ Decide on a process for handling purchases of materials, merchandise and services that will minimize the financial risks for the designer. If the amount of the purchases is too large for the designer to risk, either have the client purchase directly or have the designer act as the client's agent.

❏ Disclaim the designer's liability and responsibility for defects in materials and production, lateness, negligence, and the failures of suppliers and other professionals working on the project. The designer can agree to assign any rights against suppliers and contractors to the client so that the client can pursue any claims.

❏ Specify that the designer will bear no liability with respect to changed circumstances, such as price increases or other changes regarding purchases of merchandise.

❏ The designer should determine if he or she will or will not prepare purchase orders for purchases of merchandise, materials, and services. The advantages of purchase orders are that they provide a documented trail of the designer's expenditures, as well as documented approvals by the client. The client should approve plans, purchase orders, and similar documents and return a signed copy to the designer.

❏ If the designer is not preparing purchase orders, require that the client provide the designer with copies of all purchase orders and invoices. This is especially needed if the designer's fee is based on the amount of expenditures.

❏ Make clear that it is the client's responsibility to pay sales tax, shipping, handling, and related charges, even if these charges do not appear on the invoices.

❏ Specify the responsibilities of the client, such as cooperating, providing information, arranging interviews, providing access to the project site, approving documents promptly, facilitating communications between the designer and others working on the project, and if necessary, designating a liaison as the designer's regular contact.

❏ Make the client responsible for the costs and accuracy of any materials that the client is to provide, and indicate that the designer will be held harmless for relying on the accuracy of these materials.

❏ Require that the designer be given prompt written notice of any deviations or other problems of which the client is aware surrounding the project.

❏ Determine the client's responsibility with respect to receiving, inspecting, and storing deliveries, and try to minimize the designer's role regarding deliveries.

❏ Negotiate the method of remuneration for the designer that is most comfortable in view of the designer's customary practices and current financial circumstances. The designer should always try to obtain a nonrefundable fee from the client to start the project.

❏ In determining the method and amount of the designer's fee, carefully consider special circumstances that may require additional preparation, increased delivery costs, greater coordination of other professionals working on the project, and other factors which will use the designer's time and resources.

❏ If the designer's fee is based on expenditures, specify whether any items are to be excluded, such as those supplied by other professionals or by the client. If the client already owns certain items that will be used in the design project, attach a list of these items to the contract so that it is clear only these items are exempt from the design fee.

❏ If the designer is being paid a flat fee, indicate that the designer will be paid additional amounts when the project extends beyond a certain date.

❏ Consider having the designer's fee adjusted for inflation where the project is to take place over a long period of time, such as several years.

❏ If the client asks for services beyond the scope of the designer's work on the project, decide how the designer will be paid for these services.

❏ Determine how many revisions to the project are allowed by the client and what additional fees will be charged for further revisions.

❏ Require that the client reimburse the designer for all expenses connected to the project. Decide if there will be a markup on these expenses when billed to the client, and how nonlocal travel expenses will be billed.

❏ Decide whether an advance will be paid by the client against expenses to protect the designer from having to finance large expenditures.

❏ Negotiate a reasonable but prompt time period for the client to pay the designer's invoices after receiving them. Thirty to sixty days is customary. Give an interest rate to be charged on late payments by the client.

❏ Make sure that the client will reimburse the designer for litigation costs and attorney's fees if the designer is the winning party in a lawsuit brought pursuant to the agreement.

❏ Give a term for the industrial design agreement which will be enough time for the designer to complete the work.

❏ Consider whether to have automatic renewals to prevent the agreement from inadvertently terminating before the project is complete.

❏ Indicate what causes are grounds for termination, and provide that prior notice is required for termination for cause.

❏ Indicate the amount of time that will be allowed to cure a breach of the agreement.

❏ Allow the designer to terminate the contract if the project is postponed for longer than a certain period of time.

❏ Reserve all rights in the contract in the event of termination so that the designer keeps his or her contract and other remedies.

❏ In the event of termination, make the client responsible for the designer's fees, as well as any purchases and expenses already incurred by the designer.

❏ Negotiate that the designer will have ownership of the design to give the designer maximum control over the use of the design. In some cases, in particular where large design fees are involved, the client may insist on having ownership rights in the design. Try to limit the client's ownership interest to only the final design that is selected and approved by the client for use in the project.

❑ Require that the client retain and pay professional consultants, such as draftsmen, architects, and engineers.

❑ The designer should never provide services for which the designer is not licensed.

❑ Get the client to consent to the designer's promotional use of the project, and determine how this use will be governed.

❑ Require that credit be given to the designer if the client documents and publicizes the project.

❑ Make clear that the designer is an independent contract and that there is no employment, joint venture, or partnership relationship between the client and the designer.

❑ Indicate that the designer can assign any payments due under the agreement, but

otherwise the agreement cannot be assigned by either party without the other party's written consent.

❑ Decide whether arbitration is an appropriate mechanism for resolving disputes under the agreement, and describe how the arbitration process will be handled.

❑ Specify that the agreement shall be binding upon the designer and the client and their heirs, successors, assigns, and the like. Indicate that the agreement is the complete understanding between the parties, and the agreement cannot be modified except by a written document signed by the designer and the client.

❑ Include any additional terms and conditions which are reasonable and beneficial in clarifying and governing the agreement.

❑ Compare paragraphs 14–18 with the standard provisions in the introduction.

Industrial Designer–Client Agreement

AGREEMENT, dated the _____ day of _____, 20 _____, between _____ (hereinafter referred to as the "Client"), whose address is _____ and _____ (hereinafter referred to as the "Designer"), whose address is _____ with respect to an industrial design project (hereinafter referred to as the "Project").

WHEREAS, Designer is a professional industrial designer of good standing;

WHEREAS, Client wishes the Designer to design the Project described more fully herein; and

WHEREAS, Designer wishes to design this Project;

NOW, THEREFORE, in consideration of the foregoing premises and the mutual covenants hereinafter set forth and other valuable considerations, the parties hereto agree as follows:

1. **Description.** The Designer agrees to design the Project in a professional and workmanlike manner in accordance with:

❏ the following plan.
❏ the plan attached hereto as Schedule A and made part of this agreement.

Project description

_____
_____
_____

Scope of work to be performed by Designer

_____
_____
_____

Scope of Designer's work set forth in project phases:
Design Research/Analysis

_____
_____
_____

Design Concepts

_____
_____
_____

Conceptual Drawings

_____
_____
_____

Models/Prototypes

_____

_____

_____

Molds/Manufactured Parts

_____

_____

_____

Assembled Product

_____

_____

_____

Packaging

_____

_____

_____

Contract Documentation/Administration:

❏ If this box is checked, the Designer shall prepare an estimated budget to include with the Design Concept for presentation to the Client, but the Client acknowledges that this budget is subject to change and is not a guarantee on the part of the Designer with respect to the prices contained therein.

As the work is in progress, the Designer shall be available to meet with the Client with the following frequency to discuss the status of the work with the Client, and to consult with the Client with respect to whether what is being designed, delivered or manufactured is in conformity with specifications and of suitable quality. However, the quality and supervision of product or manufacturing shall be the responsibility of the suppliers or contractors.

Services to be rendered by the Designer in addition to those described above:

_____

_____

_____

2. Schedule. The work to be performed by the Designer under this agreement shall commence:
❏ on the _____ day of _____, 20_____
❏ if the Client is to provide reference, layouts, specifications, or other materials specified here, after the Client has provided same to the Designer.

The Designer shall make reasonable efforts to progress the Project and Designer shall:
❏ complete the Project on the _____ day of _____, 20_____
❏ conform to the following schedule: _____.

The Client understands that delays on the part of the Client or suppliers may delay performance of the Designer's duties, and the Designer's time to perform shall be extended if such delays occur. In addition, if either party hereto is unable to perform any of its obligations hereunder by reason of fire or other casualty, strike, act, or order of a public authority, act of God, or other cause beyond the control of such party, then such party shall be excused from such performance during the pendency of such cause.

3. Purchases shall be handled in the following manner:

❏ The Client shall arrange and pay for purchases of merchandise, materials and production.

❏ The Designer shall arrange and commit to purchases of merchandise, materials, and production as the agent of the Client, and the Client shall make payment directly to the suppliers or contractors. If the Designer is acting as agent for the Client, the Client shall approve all purchases by signing a written authorization.

❏ Other arrangement

The Designer shall [shall not] prepare purchase orders for purchases of merchandise, materials, or production and shall advise Client as to acceptability, but shall have no liability for the lateness, malfeasance, negligence, or failure of suppliers or contractors to perform their responsibilities and duties. In the event that, after the Client's approval for purchases of merchandise, materials or production, changed circumstances cause an increase in price or other change with respect to any such purchases of merchandise, materials, or production, the Designer shall notify the Client in writing, but shall bear no liability with respect to the changed circumstances, and the Client shall be fully responsible with respect to the purchases of merchandise, materials or production. The Designer makes no warranties or guarantees as to merchandise, materials, or production, but will assign to the Client any rights the Designer may have against suppliers or contractors, and the Client may pursue claims against such suppliers or contractors at its own expense.

The Client shall make certain that the Designer receives copies of all purchase orders and invoices.

The Client shall be responsible for the payment of sales tax, packing, shipping, and any related charges on such purchases.

4. Approvals by Client. On the Designer's request, the Client shall approve plans, drawings, renderings, purchase orders, and similar documents by returning a signed copy of each such document or a signed authorization referencing such documents to the Designer.

5. Client Responsibilities. The Client shall cooperate throughout the Project by promptly providing the Designer with necessary information; arranging any interviews that may be needed; making access available to the project site; giving prompt attention to documents to review and requested approvals; facilitating communications between the Designer and other professionals, such as engineers, suppliers, and manufacturers whom the Client has retained; and, if necessary, designating the following person to act as liaison with the Designer. If the Client is to provide specifications, plans, drawings, materials, or related information, this shall be at the Client's expense, and the Designer shall be held harmless for relying on the accuracy of what the Client has provided. If at any time the Client has knowledge of a deviation from specifications or other problem with the Project, the Client shall promptly give notice in writing to the Designer. The Client shall be responsible for receiving, inspecting, and storing all deliveries, except as follows

_____

_____

_____

6. Remuneration. Client agrees to pay the Designer on the following basis, as selected by a check mark in the appropriate box or boxes:

❏ Hourly or per diem rate. The Designer shall bill its usual hourly or per diem rates as follows: _____. The Client shall pay a nonrefundable design fee of $_____ on the signing of this Agreement. Thereafter, the Client shall pay a monthly retainer on the first day of each month in the amount of $_____ until _____. The Designer shall render billings on a _____basis, applying the design fee and retainer payments thereto, and the Client shall pay any balance due on the billings within ___ days of receipt.

❏ Flat fee plus a percentage of costs. The Designer's compensation shall be a nonrefundable design fee of $_____ paid by the Client on the signing of this Agreement, plus an additional markup of ____ percent of the expenditures for merchandise and materials and ____ percent of the expenditures for production, except that the following budget items shall not be included in this calculation:

❏ Flat fee. The Designer's compensation shall be a design fee of $_____, which shall start with a nonrefundable payment of $_____ by the Client, with the balance paid in installments of $_____ according to the following schedule:

If the Project continues beyond _____, an additional fee of $_____ shall be paid each month until completion.

In the event that the Designer's remuneration is based, in whole or in part, on the amount expended for the Project, all purchases of merchandise, materials, and production for the Project shall be included for the purpose of computing remuneration due the Designer, except for the following exclusions

_____

_____.

In addition, the Client may append to this Agreement a list of items owned by the Client prior to commencement of the Project and use these items without any additional fee being charged by the Designer.

❏ If an estimated budget is required by the Client, an additional fee of $_____ shall be charged for its preparation.

In the event that the Client requests design services beyond the scope of work for this Project and the Designer is able to accommodate the Client's request, the Designer shall bill for such additional services as follows

_____

_____.

7. Revisions. During the development of the Project, the Designer shall make a reasonable amount of revisions requested by the Client without additional charge, but if the revisions are requested after approvals by the Client, an additional fee shall be charged as follows

_____

_____.

8. Expenses. In addition to the payments pursuant to Paragraphs 6 and 7, the Client agrees to reimburse the Designer for all expenses connected to the Project, including but not limited to messengers, long-distance telephone calls, overnight deliveries, and local travel expenses. These expenses shall be marked up ____ percent by the Designer when billed to the Client. In the event that travel beyond the local area is required, the expenses for this travel shall be billed as follows

_____

_____.

At the time of signing this Agreement, the Client shall pay the Designer $_____ as a nonrefundable advance against expenses. If the advance exceeds expenses incurred, the credit balance shall be used to reduce the fee payable or, if the fee has been fully paid, shall be reimbursed to the Client. Expenses shall in no event include any portion of Designer's overhead.

9.   Payment. The Client agrees to pay the Designer within _____ days of receipt of the Designer's billings for purchases, remuneration, or expenses. Overdue payments shall be subject to interest charges of _____ percent monthly. In the event that the Designer is the winning party in a lawsuit brought pursuant to this Agreement, the Client shall reimburse the Designer for the costs of the lawsuit, including attorney's fees.

10.   Term and Termination. This Agreement shall have a term that expires on _____, 20___. The term shall automatically renew for additional _____ periods unless notice of termination is given either by party thirty (30) calendar days in advance of the renewal commencement. In addition, this Agreement may be terminated at any time for cause by either party notifying the other party in writing of that party's breach of the Agreement and giving ten (10) business days for a cure, after which the notifying party may terminate if there has been no cure of the breach. Causes for termination shall include, but not be limited to, failure to perform any duty pursuant to this Agreement in a timely manner and postponements of the Project for more than ____ business days in total. While reserving all other rights under this Agreement, in the event that the Project is terminated, the Designer shall have the right to be paid by the Client through the date of termination for the Designer's work, for any purchases by the Designer of merchandise, materials, and production pursuant to purchase orders approved by the Client, and for the Designer's expenses.

11.   Ownership of Design. The ownership of designs created by the Designer pursuant to this agreement shall be as follows:

❑ The Designer shall retain ownership of the designs, including any drawings, renderings, sketches, samples, or other materials prepared by the Designer during the course of the Project. The Designer's ownership shall include any copyrights, trademarks, patents, or other proprietary rights existing in the design. The Client shall not use the designs for additions to this Project or for any other project without the permission of the Designer and appropriate compensation.

❑ The Client shall retain ownership of any design in its final form that is selected and approved for use by the Client for the Project. All other designs, concepts and ideas prepared by the Designer in the course of the Project shall be owned by and be the exclusive property of the Designer.

12.   Consultants. If outside consultants, including but not limited to architects, structural engineers, mechanical engineers, acoustical engineers, and lighting designers, are needed for the Project, they shall be retained and paid for by the Client, and the Designer shall cooperate fully with these consultants. Such consultants shall be responsible for code compliance in the various areas of their expertise.

13.   Publicity. The Designer shall have the right to document the Project in progress and when completed, by photography or other means, which the Designer may use for portfolio, brochure, public display, and similar publicity purposes. The name of the Client and location of the Project may be used in connection with the documentation, unless specified to the contrary as follows

_____

_____.

If the Designer chooses to document the Project, the Designer shall pay the costs of documentation. In addition, if the Client documents the Project, the Designer shall be given credit as the designer for the Project if the Client's documentation is released to the public.

14.   Relationship of Parties. Both parties agree that the Designer is an independent contractor. This Agreement is not an employment agreement, nor does it constitute a joint venture or partnership between the Designer and Client. Nothing contained herein shall be construed to be inconsistent with this independent contractor relationship.

15.   Assignment. This Agreement may not be assigned by either party without the written consent of the other party hereto, except that the Designer may assign payments due hereunder to other parties.

16.   Arbitration. All disputes arising under this Agreement shall be submitted to binding arbitration before _____ in the following location _____ and settled in accordance with the rules of the American Arbitration Association. Judgment upon the arbitration award may be entered in any court having jurisdiction thereof. Disputes in which the amount at issue is less than $_____ shall not be subject to this arbitration provision.

17.   Additional Terms and Conditions:

_____

_____

_____

_____

_____

18.   Miscellany. This Agreement shall be binding upon the parties hereto, their heirs, successors, assigns, and personal representatives. This Agreement constitutes the entire understanding between the parties. Its terms can be modified only by an instrument in writing signed by both parties. Notices shall be sent by certified mail or traceable overnight delivery to the parties at the addresses shown herein, and notification of any change of address shall be given prior to that change of address taking effect. A waiver of a breach of any of the provisions of this Agreement shall not be construed as a continuing waiver of other breaches of the same or other provisions hereof. This Agreement shall be governed by the laws of the State of _____.

IN WITNESS WHEREOF, the parties hereto have signed this Agreement as of the date first set forth above.

Designer _____   Client _____
             Company Name   Company Name

By _____   By _____
     Authorized Signatory, Title   Authorized Signatory, Title

Industrial Design Contract Summary Sheet

Client
_____

Address
_____
_____

Telephone _____   Fax _____
Cell phone _____   E-mail _____

Contact at Client
_____

Representative for Designer
_____

Description of Project
_____

Full description appears in ❏ the agreement ❏ the schedule attached to the agreement

When was the agreement entered into? _____

When does the agreement commence? _____

What is the schedule for the Project? _____

Is there a projected completion date? _____

How will purchases of merchandise, materials, or production be handled in terms of which party is paying, which party is preparing purchase orders, substitutions, exclusions, and what deposits or working funds will be made available?
_____

How will remuneration be handled, including requests for work beyond the scope of the Project?
_____

Has an estimated budget been requested for inclusion with the presentation of the design concept?   ❏ Yes   ❏ No

If such estimated budget has been requested, will an additional fee be charged and, if so, how will it be computed?
_____

When will revisions be billable? _____

What expenses are billable? _____

How long does the client have to pay billings after receipt of invoices? _____

What interest rate can be charged on overdue invoices? _____

What is the term of the agreement? _____

Has the client placed any restrictions on publicizing the project? _____

What is the position of the person signing the agreement for the Client? _____

Does the agreement include the Designer's standard provisions as follows:

That the quality and supervision of merchandise or construction shall be the responsibility of the suppliers or contractors?   ❑ Yes   ❑ No

That delays caused by the Client, suppliers, contractors, or acts of God, etc. shall extend the designer's time to perform?   ❑ Yes   ❑ No

The Designer shall have no liability for the lateness, malfeasance, negligence, or failure of suppliers or contractors to perform their responsibilities and duties?   ❑ Yes   ❑ No

That the Designer shall have no liability if changed circumstances cause a change in price or other change after Client's approval for purchases of merchandise or construction?   ❑ Yes   ❑ No

That the Designer makes no warranties or guarantees as to merchandise, materials or production, but will assign to Client any rights Designer may have?   ❑ Yes   ❑ No

The Client shall be responsible for the payment of sales tax, packing, shipping, and any related charges on such purchases?   ❑ Yes   ❑ No

That the Client must sign approvals on the Designer's request?   ❑ Yes   ❑ No

That the Client must meet the usual client responsibilities?   ❑ Yes   ❑ No

Does the Designer own all rights in the designs?   ❑ Yes   ❑ No

Is the Client responsible for retaining and paying other professional consultants?   ❑ Yes   ❑ No

Is there an arbitration provision?   ❑ Yes   ❑ No

Is there a standard "Miscellany" provision?   ❑ Yes   ❑ No

Has any standard provision been altered, and have any additional terms been added to the agreement?

_____

Person completing this Contract Summary Sheet _____

Date completed _____

Commission Contract for Art or Photography

Industrial designers deal with artists in a variety of ways. A project may require the purchase of existing artworks or the commissioning of new ones. Documentation of projects may require illustration or photography. Corporate brochures for clients may use both artwork and documentation to portray the corporate design values and culture. The design firm may also want to use art in its own brochures and promotions.

These artworks must satisfy not only the designer, but the client as well. While artworks occasionally may be obtained from stock libraries, it is more likely that the designer will assign a freelance artist to create the needed works. To ensure a greater likelihood of satisfaction, the specifications for the artworks must be as clear and detailed as possible.

Of course, there must be agreement as to the fee and what is purchased for the fee. Fine artists are primarily interested in selling physical artworks. Most illustrators and photographers seek to sell only limited reproduction rights. If greater rights are purchased, they ask for a higher fee. If the designer (or client) is sensitive to this, the best approach may be to ask for limited rights. This should prevent paying for usage rights that are never exploited.

On the other hand, the designer must obtain all of the rights which his or her client needs. In the first instance, the designer must consider what rights will be transferred to the client. Rights can be limited in many ways, including the duration of use, geographic area of use, type of product or publication, title of the product or publication, and whether the use is exclusive or nonexclusive. Even in a transaction in which the designer is acquiring physical artworks for a client, there should be a consideration of usage rights. This is because every transfer of an artwork raises two questions: 1) who owns the physical artwork, which may have more than one manifestation (from color printouts from a computer to glossy posters to paint on canvas); and 2) who owns which rights of copyright (i.e., the right to make reproductions), and are these rights exclusive or nonexclusive (if they are exclusive, only the holder of the rights can exercise them).

If a fine artist sells a physical artwork, the expectation would be for the artist to retain the rights of copyright. If the designer or the client needs usage rights, these would probably be nonexclusive and limited in scope.

On the other hand, the designer or the client who licenses illustration or photography may want all rights of copyright. This would mean the designer or client could use the work in any conceivable way. However, on closer examination, it often develops that all rights are not needed. Rather, there is a desire to prevent competitors from using the illustration or photography. Another approach would be for the artist to promise by contract that no use will be made of the artworks in certain markets without first obtaining the written consent of the designer or client. Yet another would be to agree that the client has exclusive rights in those markets where the client faces competitors, but that the client will not unreasonably withhold from the artist the right to resell the artwork in a noncompetitive way. In any case, the designer must sometimes act as an intermediary—and, perhaps, as a mediator of sorts—between the demands of the client and the desire of the artist to retain rights and earn more money for greater usage.

Expenses can be a significant aspect of the cost of illustration and, especially, photography. The designer has to know the likely range of these expenses, perhaps by setting a maximum budget to be spent. If the designer requires changes, revisions, or reshoots, this will also add to the expense. Here the designer has to be careful not to be caught in a squeeze between a client with a limited budget and an artwork cost that exceeds that budget because of changes. It would be unusual to pay for expenses when purchasing a physical artwork from a fine artist, since the expenses are usually included in the fee.

In fact, there is a fundamental issue about payment. Fees and expenses for photographers or illustrators can be substantial. Should the designer become liable for such sums at all? This same issue is present in fabrication contracts and, generally, whenever the designer becomes legally obligated to a supplier for items intended for a client. While designers often commit to pay suppliers, if the costs are very large, it may be better to have the client contract directly with the supplier and pay the supplier directly. If the designer is billing on a cost-plus basis, the designer will have to have a method to charge the markup in situations in which the client does pay the fee directly.

The artist must also work on schedule. Failure to do this should be a reason for the designer to terminate the contract.

A number of professional references will aid the designer in dealing with photographers or illustrators. These include *Pricing and Ethical Guidelines* (Graphic Artists Guild, distributed by North Light Books) and *Pricing Photography* by Michal Heron and David MacTavish (Allworth Press).

Filling in the Form

Fill in the date and the names and addresses for the artist, illustrator, or photographer, as well as the designer. In paragraph 1, give the project title and description, a description of the artworks to be created, specifications for the artworks, and any other services the artist will perform. In paragraph 2, specify the amount of time the artist has to complete the assignment, including any procedures to review work in progress. In Paragraph 3, fill in which party will own the physical artworks as well as any preliminary or other materials (such as outtakes from a photography shoot).

If reproduction rights are being licensed, in paragraph 4, fill in the nature of the use, the name of the product or publication, any limitations on the geographic extent or duration of the grant of rights, and whether the use is exclusive or nonexclusive. In paragraph 5, fill in the amount of the fee, including a computation method if the fee is variable. In paragraph 6, fill in the maximum amount which the artist is allowed to bill for expenses. In paragraph 8, indicate how revisions or reshoots will be charged for by the artist. In paragraph 9, indicate whether or not the artist shall receive authorship credit.

In paragraph 13, fill in who will arbitrate, the place of arbitration, and the maximum amount that can be sued for in small claims court. State the term of the agreement in paragraph 15. In paragraph 16, specify which state's laws will govern the contract. Both parties should then sign the contract.

Negotiation Checklist

❏ Describe the assignment in whatever detail is required, attaching another sheet to the contract if necessary. It is very important to determine exactly what the artist is agreeing to do, including any services beyond creating the artworks (such as meetings with clients or proofing). (Paragraph 1)

❏ Give specifications in detail, such as black-and-white or color (and number of colors, if appropriate), number of artworks, form in which the artworks are to be delivered, and whatever else is known at the time of signing the agreement. (Paragraph 1)

❏ Approve the work-in-progress at as many stages as possible. (Paragraph 2)

❏ Give a due date for the work to be completed, as well as due dates for each approval stage. (Paragraph 2)

❏ If the designer is to provide reference materials, the due date can be stated as a number of days after the artist's receipt of these materials. (Paragraph 2)

❏ If even a short delay would cause serious problems, make time of the essence.

❏ State that illness or other delays beyond the control of the artist will extend the due date, but only up to a limited number of days.

❏ Specify who owns the physical artworks in whatever form has been specified. This might be a painting, a sculpture, transparencies, negatives, Photoshop files, and so on. (Paragraph 3)

❏ Unless there is a special reason to obtain ownership of preliminary materials used to create the artwork, the ownership of these materials would be retained by the artist. This would include the photographer's outtakes. (Paragraph 3)

❏ Be certain the grant of rights encompasses all the rights needed by the designer and, of course, by the designer's client. (Paragraph 4)

❏ State that the grant of rights extends to the client or, depending on the designer's contract with the client, gives the designer the right to assign rights to the client. (Paragraph 4)

❏ If it is likely a certain type of additional usage will be made, the amount of the reuse fee can be specified, the reuse fee can be expressed as a percentage of the original fee, or the original fee can be increased and the grant of rights expanded. Another approach would be to seek all rights, but illustrators or photographers object to selling rights that may not be used and for which nothing is presumably being paid. In any case, the fact that usage fees must be paid (and permission obtained) for uses beyond the grant of rights should be kept in mind.

❏ Specify the fee. This would also cover any possible variations in the fee, such as a greater fee for the use of more artworks or for a greater media exposure than originally planned. (Paragraph 5)

❏ Determine whether sales tax must be paid. Many states provide that the license of a right of copyright does not transfer tangible property and is not taxable (assuming the physical illustrations or photographs are returned to the creator). However, the sales tax laws vary from state to state and must be checked for the particular state involved.

❏ Any expenses that the designer will reimburse to the artist should be specified to avoid misunderstandings. Some illustrators include expenses in their fee (especially if the expenses are minimal), and the designer can certainly ask that this be done, but many illustrators and virtually all photographers bill separately for expenses. (Paragraph 6)

❏ If expenses are to be reimbursed, consider putting a maximum amount on how much will be reimbursed. Any expenses beyond this amount would have to be absorbed by the artist. This makes sense if the cap is based on an estimate provided by the artist. After receiving an itemized estimate of expenses from the artist, the designer may wish to attach this to the contract and state that expenses shall not exceed the estimate by more than 10 percent without the consent of the designer. (Paragraph 6)

❏ Determine whether the artist marks up expenses, such as billing 15 to 20 percent of the expenses as an additional charge. If expenses are going to be marked up, this should be stated. (Paragraph 6)

❏ If expenses will be significant, consider whether an advance against expenses is justified. If an advance against expenses is given, it should certainly have to be repaid if the expenses are never incurred.

❏ State that payment shall be made within a certain number of days after delivery of the finished artworks, usually within thirty days after such delivery. Obviously, this should be after the date when payment will be received from the client, unless the designer is willing to bear the negative cash flow. (Paragraph 7)

❏ Deal with the issue of payment for work-in-progress that is postponed, but not cancelled. A pro rata billing might be appropriate to handle this. (Paragraph 7)

❏ The fee for cancellation of the assignment should be specified. The designer should have the right to stop work on the project without being liable for more than the work done to date by the artist, unless special circumstances have caused the artist to have other losses. Such losses might, for example, be caused by cancellation on such short notice that a photographer is unable to schedule other work. (Paragraph 7)

❑ Specify any advances to be paid against the fee. A schedule of payments might be necessary for an extensive job, in which case the designer might also want advances from the client.

❑ Revisions or reshoots can be a problem. The artist should be given the first opportunity to make revisions or do reshoots, after which the designer should be able to change to another artist. (Paragraph 8)

❑ If revisions or reshoots are the fault of the artist, no additional fee should be charged. However, if the designer changes the nature of the assignment, additional fees will be payable. Again, if the designer is making changes because of changes by the client, the designer's contract with the client will have to provide for additional payments. (Paragraph 8)

❑ If the artist is to receive authorship credit, the designer may allow the artist to remove his or her name if changes are done by someone else. (Paragraph 8)

❑ With respect to revisions or the assignment itself, the designer should seek to avoid forcing the artist to rush or work unusual hours, since the fees for work under such stress may be higher.

❑ Document any changes in the assignment in writing, since there may later be a question as to whether the changes were executed accurately and whether they came within the initial description of the project. Paragraph 16 requires that all modifications to the agreement be written. Form 10, the Work Change Order, can be used as necessary to document changes.

❑ State whether the artist will receive name credit with the image. (Paragraph 9)

❑ State whether copyright notice for the photographs or illustrations will appear in the name of the photographer or illustrator when the design is published. (Paragraph 10)

❑ The artist must obtain releases with respect both to using copyrighted work or, in some cases, using art created by other artists. Releases for models may also be necessary. Such releases should protect both the designer and the designer's client. (Paragraph 11)

❑ The designer may want a warranty and indemnity provision, in which the artist states the work is not a copyright infringement and not libelous and agrees to pay for the designer's damages and attorney's fees if this is not true. Such a warranty should not extend to materials provided by the designer or client for use in the artwork. (Paragraph 12)

❑ Include a provision for arbitration, except for amounts that can be sued for in small claims court. (Paragraph 13)

❑ Allow the artist the right to assign money payable under the contract, unless there is a particular reason not to do so. (Paragraph 14)

❑ Give the designer the right to assign the contract or rights under the contract. The designer will want to assign rights to the client. (Paragraph 14)

❑ Specify a short term for the agreement. (Paragraph 15)

❑ Allow the designer to terminate if the artist does not meet the project's specifications, falls behind schedule, or becomes insolvent. (Paragraph 15)

❑ Compare the standard provisions in the introductory pages with Paragraph 16.

Other Provisions That Can Be Added to Form 37

❑ Noncompetition. If a client is concerned about competitors using the same art or having a similar look, one solution is to insist on an all-rights contract. The artist would have no right to reuse the work at all. A less extreme solution is to have a noncompetition provision, although even this can be objectionable, since the artist cannot risk his or her livelihood by agreeing not to work in a particular style. In any case, a noncompetition provision might read as follows:

Noncompetition. The Artist agrees not to make or permit any use of the Artworks or similar artworks that would compete with or impair the use of the Image by the Designer or its client. The Artist shall submit any proposed uses of the Artworks or similar artworks to the Designer for approval, which approval shall not be unreasonably withheld.

Commission Contract for Art or Photography

AGREEMENT, dated the _____ day of _____, 20 _____, between _____ (hereinafter referred to as the "Artist"), whose address is _____ and _____ _____ (hereinafter referred to as the "Designer"), whose address is _____ _____ with respect to the creation of certain artworks (hereinafter referred to as the "Artworks").

WHEREAS, Artist is a professional artist, illustrator, or photographer of good standing;

WHEREAS, Designer wishes the Artist to create the Artworks described more fully herein; and

WHEREAS, Artist wishes to create such Artworks pursuant to this Agreement;

NOW, THEREFORE, in consideration of the foregoing premises and the mutual covenants hereinafter set forth and other valuable considerations, the parties hereto agree as follows:

1. **Description.** The Artist agrees to create the Artworks in accordance with the following specifications:

Project title and description of the Artworks

_____
_____
_____

Other specifications

_____
_____
_____

Other services to be rendered by the Artist

_____
_____
_____

2. **Due Date.** The Artist agrees to deliver the Artworks within _____ days after the signing of this Agreement or, if the Designer is to provide reference, layouts, or specifications, after the Designer has provided same to the Artist. If the Designer is to review and approve the work in progress, specify the details here

_____
_____
_____

3. **Ownership of Physical Artworks.** The ownership of the physical Artworks as specified in Paragraph 1 shall be the property of _____. Sketches and any other materials created in the process of making the finished Artworks shall remain the property of the Artist, unless indicated to the contrary here

_____

4.   Grant of Rights. The Artist grants to the Designer the right to display the Artworks to the public, as well as the following additional usage rights:

For use as _____

For the product or publication named _____

These rights shall be worldwide and for the full life of the copyright and any renewals thereof unless specified to the contrary here

_____

These rights are
❑ exclusive
❑ nonexclusive

5.   Fee. The Designer agrees to pay the following price: $_____. If usage rights are granted and the fee varies with the amount or nature of usage, the fee shall be computed as follows

_____

_____

_____

6.   Expenses. The Designer agrees to reimburse the Artist for expenses incurred in creating the Artworks, provided that such expenses shall be itemized and supported by invoices, shall not be marked up, and shall not exceed $_____ in total.

7.   Payment. The Designer agrees to pay the Artist within thirty days of the date of the Artist's billing, which shall be dated as of the date of delivery of the Artworks. In the event that work is postponed or cancelled at the request of the Designer, the Artist shall have the right to bill and be paid pro rata for work completed through the date of that request, but the Designer shall have no further liability hereunder.

8.   Revisions or Reshoots. The Artist shall be given the first opportunity to make any revisions or reshoots requested by the Designer. If the revisions or reshoots are not due to any fault on the part of the Artist, an additional fee shall be charged as follows

_____

_____.

If the Artist objects to any revisions to be made by the Designer, the Artist shall have the right to have any authorship credit and copyright notice in his or her name removed from the Artworks.

9.   Authorship Credit. Authorship credit in the name of the Artist ❑ shall ❑ shall not accompany the Artworks when reproduced.

10.   Copyright Notice. Copyright notice in the name of the Artist ❑ shall ❑ shall not accompany the Artworks when reproduced.

11.   Releases. The Artist agrees to obtain releases for any art, photography, or other copyrighted materials to be incorporated by the Artist into the Artworks as well as for any models who will be portrayed in the Artworks.

12.   Warranty and Indemnity. The Artist warrants and represents that he or she is the sole creator of the Artworks and owns all rights granted under this Agreement, that the Artworks are an original creation (except for materials obtained with the written permission of others or materials from the public domain), that the Artworks do not infringe any other person's copyrights or rights of literary property, nor do they violate the rights of privacy of, or libel, other persons. The Artist agrees to indemnify and hold harmless the Designer against any claims, judgments, court costs, attorney's fees, and other expenses arising from any alleged or actual breach of this warranty.

13.   Arbitration. All disputes arising under this Agreement shall be submitted to binding arbitration before _____ in the following location _____ and settled in accordance with the rules of the American Arbitration Association. Judgment upon the arbitration award may be entered in any court having jurisdiction thereof. Disputes in which the amount at issue is less than $_____ shall not be subject to this arbitration provision.

14.   Assignment. The Designer shall have the right to assign any or all of its rights and obligations pursuant to this Agreement. The Artist shall have the right to assign monies due to him or her under the terms of this Agreement, but shall not make any other assignments hereunder.

15.   Term and Termination. This Agreement shall have a term ending _____ months after payment pursuant to Paragraph 6. The Designer may terminate this Agreement at any time prior to the Artist's commencement of work and may terminate thereafter if the Artist fails to adhere to the specifications or schedule for the Artworks. This Agreement shall also terminate in the event of the Artist's bankruptcy or insolvency. The rights and obligations of the parties pursuant to Paragraphs 3, 4, 9, 10, 11, 12, 13, and 14 shall survive termination of this Agreement.

16.   Miscellany. This Agreement constitutes the entire understanding between the parties. Its terms can be modified only by an instrument in writing signed by both parties. A waiver of a breach of any of the provisions of this Agreement shall not be construed as a continuing waiver of other breaches of the same or other provisions hereof. This Agreement shall be binding upon the parties hereto and their respective heirs, successors, assigns, and personal representatives. This Agreement shall be governed by the laws of the State of _____.

IN WITNESS WHEREOF, the parties hereto have signed this Agreement as of the date first set forth above.

Artist _____   Designer _____
                Company Name

By _____
            Authorized Signatory, Title

Confidentiality Agreement

Industrial designers are involved in numerous innovative and proprietary design concepts. Whether these concepts originate from the designer or they are acquired by contract, they can be valuable intellectual property assets. Depending on the concept, it may be protectable by utility or design patents, copyrights, trademarks or trade secrets. These forms of intellectual property protection are discussed later in this book, but each should be carefully explored by the industrial designer as a means to safeguard novel designs and other proprietary information.

The first step in safeguarding designs and proprietary information is to maintain confidentiality until they can be protected by other forms of intellectual property. There are several reasons for maintaining the confidentiality of an idea or concept and restricting or preventing its disclosure to others. The designer may want to keep the idea confidential to prevent someone else from stealing it before other legal protection can be obtained for it. The designer may also want to maintain the confidentiality of an idea to retain the ability to file for a patent. Although the United States gives a one-year grace period, many countries require that there be no public use, sale, or disclosure of the concept before the designer has filed a patent application covering the concept. Therefore, initially maintaining confidentiality of the idea allows the designer to protect it later for optimum commercial exploitation.

Trade secrets are usually the first way to protect proprietary information. The subject matter covered by trade secret protection is broad. A trade secret is generally defined as any formula, pattern, device, or compilation of information which is or can be used in one's business, and which gives one an opportunity to obtain an advantage over competitors who do not know or use it, and which the law protects from misappropriation by others.

Thus, almost any information used in business can be a trade secret if it is not generally known and if proper precautions are taken to maintain confidentiality. For example, trade secrets can be applied to designs, programs, techniques, customer lists, business plans, methods of doing business, and more. Trade secrets are recognized even when the information has been disclosed to contractors, suppliers, and others dealing with the trade secret holder, provided that the information is disclosed under a confidentiality agreement.

For the designer to protect and profit from his or her concepts from a practical point of view, the designer will have to disclose the concepts to other people, such as witnesses to an invention, employees, draftspeople, patent lawyers, consultants, manufacturers, and customers. Otherwise, the designer will not be able to have the design concepts tested, patented, licensed, produced, or marketed. The best approach is always to limit disclosure of the concepts to those who have a need to know in order to assist the designer in protecting and profiting from the design concepts. Disclose unpatented or unprotected concepts only after the parties to whom the disclosure is being made agree in writing to keep the concepts confidential.

When someone serves as a witness to the invention of the designer's concepts, the witnesses should be asked to sign a written confidentiality agreement. Similarly, the designer should require that employees, agents and consultants sign an appropriate confidentiality agreement. Include provisions prohibiting employees and consultants from misappropriating the designer's concepts to compete with him. Furthermore, the designer should require employees and consultants to assign to the designer any improvements they make to the designer's concepts.

When the designer submits concepts to interested companies or customers for their review and evaluation, the designer should briefly state the purpose and advantages of his or her concept, and require the companies and customers to sign a confidentiality agreement before they can obtain more detailed information. Some companies may refuse to sign a confidentiality agreement before reviewing the idea. If such a company is important for the designer to profit from the concept, the designer may have to disclose the concept even if the company is unwilling to enter into a confidentiality agreement. If the designer's concept is patented or otherwise protected, there is less risk to the designer in disclosing it. Nevertheless, when the designer discloses unprotected concepts without a confidentiality agreement, it is a good idea to mark the copies as confidential and indicate in a letter to the recipient that the concept is submitted in confidence.

If the designer discusses the concept with an attorney for patent or legal advice, the attorney is generally under a professional duty to maintain the confidentiality of privileged information. If the attorney violates this duty of confidentiality, he or she can face professional discipline and civil liability for damages caused to the designer by the unauthorized disclosure. When the designer retains an attorney, it is recommended that an attorney retainer agreement be entered into which specifically obligates the attorney to keep the concept in confidence and not use or disclose it to others without the designer's consent.

A sample confidentiality agreement is provided at the end of this chapter. The sample confidentiality agreement is adaptable for use in practically any situation where the designer will disclose proprietary information or concepts to others.

Filling in the Form

In the preamble of the confidentiality agreement form, fill in the date and the names and addresses of the designer and the recipient. In paragraph 1 describe in general terms the information that the designer will be disclosing without giving away exactly what it is. In paragraph 4 indicate the number of years that the recipient is to keep the disclosed information in confidence. Include any additional terms and conditions in paragraph 6. In paragraph 8, fill in the name of the state where the confidentiality agreement will be governed. Have both the recipient and the designer sign the confidentiality agreement, and include their names and titles.

Negotiation Checklist

❏ Describe in very general terms what the proprietary information or design concept concerns, but do not give away what is novel and innovative about them. For example, "an idea for a unique design for a chair" might be a sufficient description but it would not give away the appearance or specifications.

❏ State that the recipient is receiving the information only for the purpose of determining its interest in the commercial exploitation of the information.

❏ Limit the use that the recipient can make of the disclosed information, and restrict the recipient from making, selling, or otherwise using the information except for the purpose of evaluation under the confidentiality agreement.

❏ Make it clear that the confidentiality agreement does not grant the recipient any rights to the information disclosed by the designer.

❏ Require that the recipient keep the information confidential and not disclose it to anyone for a sufficient period of time to allow the designer to patent or control the release of it. Five to ten years is the customary period for confidentiality, but negotiate for as long a period as possible.

❏ Require that the recipient acknowledge that unauthorized disclosure of the information will cause irreparable harm to the designer.

❏ Provide for good faith negotiations toward a satisfactory agreement if the recipient wishes to use the designer's information.

❏ The confidentiality agreement should remain in effect until a satisfactory agreement regarding use of the information is entered into.

❏ Make sure that the confidentiality agreement is binding upon and inures to the benefit of the parties and their agents, successors, and assigns.

❏ State that the confidentiality agreement constitutes the entire agreement of the parties, and that it can only be amended by written agreement signed by the designer and the recipient.

❏ Decide on the state where the confidentiality agreement will be governed. This should generally be the state where the designer is located or has easy access and familiarity.

❏ Be certain that the confidentiality agreement is signed by an authorized representative of the recipient.

If the designer wishes to disclose the information despite the recipient's refusal to sign the confidentiality agreement, the designer should take a number of steps:

❏ Document the designer's concept by writing a complete description of it, including any drawings where appropriate. Have this document signed and dated by the designer and at least one independent witness who can understand the concept.

❏ Explore all means to protect the concept, such as utility patents, design patents, copyrights, and trademarks. Where protection has been granted include on the information the appropriate patent, copyright, or trademark notice. If a patent application has been filed and is under examination, mark the concept as "patent pending." These notices can serve as a deterrent against misappropriation of the information by others.

❏ Mark all information that will be disclosed as confidential and indicate that it is being submitted in confidence.

❏ If an appointment is made, the designer should confirm it in writing in advance and sign any log for visitors.

❏ After any meeting, the designer should send a letter to the recipient summarizing what happened at the meeting (including any disclosure of confidential information and any assurances that information will be kept confidential). If at all possible, have any proposal or follow-up from the recipient be in writing.

Confidentiality Agreement

AGREEMENT, dated the _____ day of _____, 20 _____, between _____ (hereinafter referred to as the "Designer"), whose address is _____ and _____ (hereinafter referred to as the "Recipient"), whose address is _____.

WHEREAS, the Designer has developed certain valuable information, concepts, ideas, designs, or products, which the Designer deems confidential (hereinafter referred to as the "Information"); and

WHEREAS, the Recipient is in the business of using such Information for its projects and wishes to review the Information; and

WHEREAS, the Designer wishes to disclose this Information to the Recipient; and

WHEREAS, the Recipient is willing not to disclose this Information, as provided in this Agreement.

NOW, THEREFORE, in consideration of the foregoing premises and the mutual covenants hereinafter set forth and other valuable considerations, the parties hereto agree as follows:

1. **Disclosure.** The Designer shall disclose to the Recipient the Information, which relates to the following:

2. **Purpose.** The Recipient agrees that this disclosure is only for the purpose of the Recipient's evaluation to determine its interest in the commercial exploitation of the Information.

3. **Limitation on Use.** The Recipient agrees not to manufacture, sell, deal in, or otherwise use or appropriate the disclosed Information in any way whatsoever except as expressly provided above, including but not limited to adaptation, imitation, redesign, or modification. Nothing contained in this Agreement shall be deemed to give Recipient any rights whatsoever in and to the Information.

4. **Confidentiality.** The Recipient understands and agrees that the unauthorized disclosure of the Information by the Recipient to others would irreparably damage the Designer. As consideration and in return for the disclosure of this Information, the Recipient shall, for a period of ___ years for the date of this agreement, keep secret and hold in confidence all such Information and treat the Information as if it were the Recipient's own proprietary property by not disclosing it to any person or entity.

5. **Good Faith Negotiations.** If, on the basis of the evaluation of the Information, the Recipient wishes to pursue the exploitation thereof, the Recipient shall enter into good faith negotiations to arrive at a mutually satisfactory agreement for these purposes. Until and unless such an agreement is entered into, this nondisclosure Agreement shall remain in force.

6. **Other Terms and Conditions.**

    _____

    _____

    _____

7. **Miscellany.** This Agreement shall be binding upon and shall inure to the benefit of the parties and their respective legal representatives, successors, and assigns.

8. Entire Agreement. This Agreement constitutes the entire agreement of the parties with respect to the Information, and it shall not be amended orally, but only by an agreement in writing signed by the Designer and the Recipient. This Agreement shall be governed by the laws of the State of _____.

IN WITNESS WHEREOF, the parties have signed this Agreement as of the date first set forth above.

Designer's Signature

_____

_____

Name/Title

Recipient's Signature

_____

_____

Name/Title

Contract with Fabricator

The industrial designer is likely to have items fabricated, either for use by the firm or a client. In either case, the designer must be protected from the risk that the item may not be produced in a satisfactory manner (especially if the client is dissatisfied with the item and refuses to pay). If the item is for a client, the safest approach would be for the designer to have the client contract directly with the fabricator or supplier, but this will be impractical if the fabricator or supplier will only work with designers. Also, the designer may prefer not to reveal the extent of the markup on certain items.

It is important that the specifications be clear, since quality is crucial. The materials to be used to make the product, the various stages for approval of what has been done, and a final sample or proof should all be required by the designer. In addition, the design firm should synchronize its approvals to the fabricator with its client's approvals. If a client is going to reject custom-made items, even if the items appear reasonably satisfactory based on the specifications, the designer must stop the production process and try to salvage whatever can be saved. The more closely the client is involved in the stages of approval, the greater the likelihood for a successful outcome.

The fabricator should be trustworthy enough to ensure proper production techniques and quality, but this is not always the case. Certainly, the designer should review samples of the fabricator's work to see if the quality is satisfactory. If the designer feels comfortable in agreeing to take the job through the completion of fabrication, the issue of the scope of the designer's duties and responsibilities must be resolved. Is the designer expected to go to the fabricator's plant, handle all contacts with the fabricator, and arrange for any corrections that are necessary? The scope of duties will be a function of whether the designer or client

is most competent in these areas. The client will often rely on the expertise of the designer, who must be able to monitor the performance of the fabricator and make the fabricator correct the errors that inevitably crop up in production.

What happens if, after the products are fabricated, the client refuses to pay the designer? The designer has a bill from the fabricator that must be paid. The typical scenario involves a client with (sometimes unreasonably) high standards who is unwilling to accept a product produced to less than the highest standards of quality. The fabricator expects the designer to accept the job, yet the designer finds the job is not acceptable to the client. Nor can the designer easily reject the job, since the fabricator has not made errors that would justify such a rejection. This nightmare can be avoided in several ways. Some designers simply refuse to handle money in relation to fabrication. They insist that the client contract directly with the fabricator, even if the designer is to render services during the fabrication process. What duties the designer performs are billed to the client on an agreed-upon basis, either as a fee, an hourly rate, or a markup. If the designer prefers to pay the fabricator, the fabricator must understand and agree to meet a quality standard consistent with the client's expectation (and the designer's agreement with the client).

The first step in dealing with a fabricator is to request a price quotation. To do this, detailed specifications must be given. It is always wise to seek more than one bid, since prices vary widely. One reason for great price variation is that fabricators have different equipment. The equipment may make the fabricator effective for one project, but not another, which is reflected in the price. Asking each fabricator about what they can do most efficiently may give helpful insights into selecting the right fabricator.

Since the specifications initially given to the fabricator are always subject to change, estimates may include variable costs for different configurations of the item, differing materials, and differing quantities. When requesting quotations, always keep the specifications identical for each fabricator. Also, find out how long a quotation will hold before the fabricator will insist on rebidding the job (which may increase the price).

Once a fabricator has given an acceptable quotation, the designer wants to know that the job will be of appropriate quality and delivered on time. It is important to check the fabricator's work at various stages before allowing the job to be produced. This is usually done by seeing work in progress and a final sample or proofs. This approval process allows the designer, in turn, to obtain approvals from the client. After approval, the fabricator is free to produce the item in conformity with the approved samples or proofs.

If a large quantity of items is being produced, the designer will have to determine whether to accept more or less than the agreed-upon quantity. These "overs" or "unders" are paid for at the marginal unit cost (and thus are relatively inexpensive). In general, overs and unders will be unacceptable to the designer whose client needs an exact quantity. If the designer is ordering large quantities and will not accept either overs or unders, this should be stated in the order. A 5 or 10 percent variation in quantity might be acceptable in a situation where the item would be marketed to the public, rather than delivered to a particular client.

The designer will certainly keep ownership over whatever materials he or she gives to the fabricator, such as computer files, models, or plans. If these materials are valuable, they should be insured and returned as soon as possible after fabrication. A more sensitive issue is ownership of the materials created by the fabricator in the course of the project. If, for whatever reason, the designer wishes to have a future job done by a different fabricator, the designer may want to own the production materials created by the fabricator. If the designer owns these production materials (such as molds or dyes), the designer might require the

fabricator to store them without charge and pay only for delivery charges if the designer decides to move the materials to another fabricator. Fabricators should agree to this, since the materials will never have to be moved if the fabricator satisfies the designer. At the same time, the designer should also be aware that the client may expect to own what both the designer and the fabricator create, so this must be resolved in a way that is consistent with the understanding between designer and fabricator. Fabricators may seek to limit their liability to the amount paid for the job. But what if an item costs $10,000 to produce, yet the designer loses a $30,000 fee because the fabricator never delivers or delivers too late? The fabricator will also prefer to deliver "F.O.B. plant," which means that the fabricator will load the job at the plant without charge, but has no responsibility after that. The designer will either arrange to pay shipping and insurance costs to the final destination or ask the fabricator to ship "C.I.F. Greenwich, Connecticut," if that is the destination. C.I.F. means that a price quotation covers the cost of the merchandise as well as insurance and freight charges to the destination. Thus, the insurance and freight charges are paid by the shipping party (in this case, the fabricator). If the fabricator does arrange this, it will no doubt want to bill an extra charge. In any case, the designer must be assured that if the job is lost or damaged in shipment, insurance funds are available to cover the loss and, if feasible, re-creation of the merchandise.

If the designer is not handling the job for the client, the client may still be very appreciative if the designer alerts the client to risks in the production process. Although Form 39 is set up for the designer to contract with and pay the fabricator, it could as easily be used between the client and the fabricator.

Filling in the Form

In the preamble, fill in the date and the names and addresses of the parties. In paragraph 1, check the appropriate box, and either fill in the specifications in paragraph 1 or add an attachment titled schedule A. For a job with several components to produce, more than one copy of

Schedule A might be used. In paragraph 2, specify the delivery date, the place of delivery, and the terms (probably F.O.B. or C.I.F.). In paragraph 3, give the price. Then, if overs and unders are acceptable, specify the price per unit for the overs and unders.

In paragraph 4, specify when payment must be made after delivery (usually thirty days for United States fabricators, although sixty days is sometimes agreed to, and, in some cases, a deposit will have to be paid). In both paragraphs 5 and 6, indicate whether the fabricator must insure the materials and, if so, for how much. In paragraph 6, also indicate which party will pay the expense of returning the materials.

In paragraph 9, specify the arbitrator, the place of arbitration, and the amount beneath which claims can be brought in small claims court. In paragraph 10, specify a term for the contract. In paragraph 11, indicate which state's laws shall govern the contract. Have both parties sign, and append a Schedule A, if necessary. The designer should either be expert enough to fill out a Schedule A or should use the assistance of a skilled production manager.

Negotiation Checklist

❑ Fill out the specifications for the product that the designer wants, including quantity, materials to be used, approvals, packing, and any other specifications. (Paragraph 1 or Schedule A)

❑ Specify a delivery date. (Paragraph 2)

❑ Specify a delivery location. (Paragraph 2)

❑ Indicate the terms of delivery, such as F.O.B. or C.I.F., and be certain the job is sufficiently insured. (Paragraph 2)

❑ State that the risk of loss is borne by the fabricator until the job is delivered according to the terms of the contract. (Paragraph 2)

❑ State that time is of the essence. The fabricator will resist this, since late delivery will be an actionable breach of contract. (Paragraph 2)

❑ Do not allow the fabricator to limit damages for nondelivery or late delivery to the purchase price of the job.

❑ State the price for the quantity ordered. (Paragraph 3)

❑ Specify whether overs or unders are acceptable and, if they are, what percentages are allowed. (Paragraph 3)

❑ Determine whether any sales or other tax must be paid, and ascertain whether this tax has been included in the price or will be an additional charge.

❑ State when payment will be made after delivery, which is usually within thirty or sixty days, although the designer may have to make a deposit if credit has not been established with the fabricator. (Paragraph 4)

❑ Do not give the fabricator a security interest in the job, which the fabricator might want until full payment has been made. Such a security interest, when perfected by filing with the appropriate government agencies, would give the fabricator a right to the materials or to any sale proceeds from the materials.

❑ State that all materials supplied by the designer remain the property of the designer and must be returned when no longer needed. (Paragraph 5)

❑ Do not give the fabricator a security interest in materials supplied by the designer.

❑ Indicate that the fabricator shall pay the expense of returning the materials supplied by the designer. (Paragraph 5)

❑ State whether materials supplied by the designer shall be insured by the fabricator and, if so, for how much. (Paragraph 5)

❑ State that all materials created by the fabricator shall be the property of the designer, must be stored without charge, and must be returned when no longer needed. (Paragraph 5)

❑ Indicate who will pay for the return to the designer of materials created by the fabricator. (Paragraph 5)

❏ State whether materials created by the fabricator and owned by the designer will be insured by the fabricator and, if they are to be insured, for how much. (Paragraph 5)

❏ Decide whether the fabricator may use other companies to do part of the production process. If the designer's trust is with a particular fabricator, this practice may be ill-advised. In any case, the designer should be familiar with the true capabilities of the fabricator. Jobbing work out may cause production delays. Also, a fabricator's bid that seems too high may be the result of the fabricator marking up work to be done by others, instead of doing that work itself.

❏ If the fabricator requests a provision to extend the delivery date in the event of war, strikes, or similar situations beyond its control, the designer should specify that after a specified period of time, the contract will terminate. This period of time might be relatively brief if the job has not yet been produced.

❏ Require approvals for all parts of the job, and hold the fabricator responsible for matching the final proofs. (Paragraph 7)

❏ Require that the fabricator meet a quality standard based on the specifications. (Paragraph 8)

❏ Do not allow the fabricator to make a blanket disclaimer of warranties, since these warranties are to protect the buyer. A warranty is a fact the buyer can rely upon, such as the fabricator's statement that a certain kind of material will be used or simply the fact that the fabricator has title to what is produced and can sell it.

❏ The designer should ascertain any extra expenses, such as charges for special work, charges for changes in the specifications after an order has been placed, charges to use materials provided by the designer on behalf of the client, charges for delays caused by the designer's tardiness in making approvals or reviewing proofs (which may be due to the client being slow), charges for having someone at the plant to approve the production as it is done, charges for samples to be airfreighted, charges for storage of merchandise, charges for shipping, and any other charges.

❏ State that disputes shall be arbitrated, but do not allow the fabricator either to have the arbitration at its sole option or to specify the location as its place of business. (Paragraph 9)

❏ Specify a short contractual term, such as a period of months. (Paragraph 10)

❏ Allow the designer to terminate without charge prior to the fabricator's commencement of work or if the fabricator fails to meet the production specifications or the production schedule. (Paragraph 10)

❏ State that the contract will terminate in the event that the fabricator becomes bankrupt or insolvent. (Paragraph 10)

❏ Decide whether the death of the designer or fabricator should terminate the agreement.

❏ Specify that the designer's right to materials it supplied or materials the fabricator created will survive termination of the contract, as will the right to arbitration. (Paragraph 10)

❏ If there is to be any charge for cancellation of an order, make certain such a charge bears a reasonable relationship to expenses actually incurred by the fabricator for that order.

❏ If work beyond the original specifications is needed, define a method or standard that the fabricator will use to bill such extra work.

❏ Do not allow the fabricator to limit the time to inspect the merchandise and complain about defects, since the designer should certainly have a reasonable amount of time to do this. What is reasonable will depend on the use to which the client will put the materials. Of course, defects should be looked for and documented in writing as soon as discovered.

❏ If the fabricator uses its own form to confirm the order, review all printed terms on the form to make certain these terms are not in conflict with the designer's forms or agreement.

❏ Review the standard provisions in the introductory pages, and compare them with paragraphs 9 and 11.

Contract with Fabricator

AGREEMENT, dated the _____ day of _____, 20 ____, between _____ (hereinafter referred to as the "Designer"), whose address is _____ and _____ (hereinafter referred to as the "Fabricator"), whose address is _____, with respect to the production of certain materials (hereinafter referred to as the "Work").

WHEREAS, the Designer has given specifications for the Work and wishes to have the Work produced in accordance with those specifications under the terms of this Agreement; and

WHEREAS, the Fabricator is in the business of producing such Work and is prepared to meet the specifications and other terms of this Agreement with respect to producing the Work;

NOW, THEREFORE, in consideration of the foregoing premises and the mutual covenants hereinafter set forth and other valuable consideration, the parties hereto agree as follows:

1. **Specifications.** The Fabricator agrees to produce the Work in accordance with [] Schedule A [] the following specifications:

Title _____
Description _____
Quantity _____

If computer files, dyes, diagrams, models or other guides are to be given to the Fabricator, specify here
_____
_____

Materials to be used in fabrication
_____
_____

Specify stages for production approvals
_____
_____

Final proofs will be in the following form, prior to production of the entire quantity of the Work
_____

Packing
_____

Other specifications
_____
_____
_____

2. **Delivery and Risk of Loss.** Fabricator agrees to deliver the order on or before _____, 20____ to the following location _____ and pursuant to the following terms _____.

The Fabricator shall be strictly liable for loss, damage, or theft of the order until delivery has been made as provided in this paragraph. Time is of the essence with respect to the delivery date.

3. Price. The price for the quantity specified in Paragraph 1 shall be $_____. Overs and unders shall not be acceptable unless specified to the contrary here _____, in which case the price shall be adjusted at the rate of $_____ per _____.

4. Payment. The price shall be payable within _____ days of delivery.

5. Ownership and Return of Supplied Materials. All camera-ready copy, artwork, film, separations, models, and any other materials supplied by the Designer to the Fabricator shall remain the exclusive property of the Designer and be returned by the Fabricator at its expense as soon as possible upon the earlier of either the producing of the Work or the Designer's request. The Fabricator shall be liable for any loss or damage to such materials from the time of receipt until the time of return receipt by the Designer. The Fabricator [] shall [] shall not insure such materials for the benefit of the Designer in the amount of $_____.

6. Ownership and Return of Commissioned Materials. All materials created by the Fabricator for the Designer, including but not limited to sketches, mechanical art, models, type, negatives, positives, flats, dyes, or plates, shall become the exclusive property of the Designer and shall be stored without expense by the Fabricator and be returned at the Designer's request. The expense of such return of materials shall be paid by the ❑ Fabricator ❑ Designer. The Fabricator shall be liable for any loss or damage to such materials from the time of creation until the time of return receipt by the Designer. The Fabricator [] shall [] shall not insure such materials for the benefit of the Designer in the amount of $_____.

7. Proofs. If proofs are requested in the specifications, the Work shall not be produced until the Designer has approved such proofs in writing. The finished copies of the Work shall match the quality of the proofs.

8. Quality. The Designer shall have the right to approve the quality of the Work based on its conformity to the specifications.

9. Arbitration. All disputes arising under this Agreement shall be submitted to binding arbitration before _____ at the following location _____ and the arbitration award may be entered for judgment in any court having jurisdiction thereof. Notwithstanding the foregoing, either party may refuse to arbitrate when the dispute is for less than $_____.

10. Term and Termination. This Agreement shall have a term ending _____ months after payment pursuant to Paragraph 4. The Designer may terminate this Agreement at any time prior to the Fabricator's commencement of work and may terminate thereafter if the Fabricator fails to adhere to the specifications or production schedule for the Work. This Agreement shall also terminate in the event of the Fabricator's bankruptcy or insolvency. The rights and obligations of the parties pursuant to Paragraphs 5, 6, and 8 shall survive termination of the Agreement.

11. Miscellany. This Agreement contains the entire understanding between the parties and may not be modified, amended, or changed except by an instrument in writing signed by both parties. A waiver of any breach of any of the provisions of this Agreement shall not be construed as a continuing waiver of other breaches of the same or other provisions hereof. This Agreement shall be binding upon the parties hereto and their respective heirs, successors, assigns, and personal representatives. This Agreement shall be interpreted under the laws of the State of _____.

IN WITNESS WHEREOF, the parties have signed this Agreement as of the date first set forth above.

Fabricator _____ Designer _____
          Company Name   Company Name

By _____ By _____
       Authorized Signatory, Title   Authorized Signatory, Title

Licensing Agreement for Industrial Designs

Licensing is a form of intellectual property exploitation that allows the designer to capitalize on the financial earning power of patents, copyrights, and trademarks covering proprietary concepts by making them available for use in exchange for some level of compensation. Licensing allows the designer to exploit the value of the design without having to give up ownership and control of the design or having to work to bring the design to commercialization. The designer licensor usually collects lump-sum payments and/or running royalties without assuming any business risks.

An important part of the licensing process is a determination of the value of the designer's concept, and this involves the examination of a number of factors. Does the designer's concept represent a technological breakthrough? Has the concept been fully developed or commercialized? What are the real and the perceived benefits of the concept? Is the concept adequately protected by patent or other intellectual property rights? These are some of the questions the designer must answer before he or she can determine the licensing potential of his or her concept.

The designer should realize that a concept not protected by a patent or otherwise will likely have little or no licensing potential. No one wants to pay for a license to a concept that the rest of the world can practice freely. Likewise, prospective licensees may not have a strong interest in a concept that has limited marketability.

It is important that the designer size up the market in which his or her concept would compete. The designer will need to assess the commercial potential and marketability of his or her concept before being able to measure its licensing value. This may require market studies and an analysis of the designer's concept in comparison to similar products as well as unrelated alternatives. The designer should look at other license agreements that he or she may have access to for determining customary royalties and conditions. When this analysis is complete, the designer is ready to examine licensing as a way to exploit his or her proprietary concept.

A license is basically a contract and requires the essential elements of an offer, acceptance and consideration. The license is a negotiated agreement that should always be in writing and signed by both the designer and the licensee. The license agreement can be simple or complex depending on the terms and conditions required by the parties. A sample license agreement for industrial designs is provided at the end of this section that can be used for most simple licensing situations. Remember that each licensing agreement is a distinct transaction with terms that are negotiated and agreed to by the parties on a case-by-case basis.

There are several prerequisites to entering into an effective license. First, the designer must have ownership of the concept and the right to license it. If the designer created the concept for his or her benefit and without any direct or indirect involvement of another party, the designer's ownership of the concept is relatively clear. Make sure that the designer's employees and consultants who worked on the concept are required by a written contract to assign all rights to the designer. However, check to be certain that the designer is not under any obligations to assign his or her rights to the concept to an employer or others.

Secondly, the license must specify what rights are being granted by the designer to the licensee. Is the license exclusive or nonexclusive? Are full rights to make use, sell, import, and distribute being granted to the licensee? Is the license restricted to a specific state, region, or country? Also, the license should clearly indicate the consideration (e.g., royalties or lump-sum payments) that is being paid by the licensee to the designer.

Under a license, the designer grants permission to the licensee to practice the concept within certain respects. The designer can give or grant the licensee permission to make, have made, use, offer for sale, sell, import, or distribute the proprietary concept and products which use or incorporate it. In granting a license, the designer can do so on an "exclusive" or "nonexclusive" basis. Where the license is exclusive, the designer agrees to license the concept only to the licensee, and it usually means that the designer gives up his or her own right to practice the concept. If large licensing fees, royalties or investments are required, the licensee is likely to press for an exclusive license. If the designer plans to use the concept in his or her own business and grants an exclusive license, the designer should expressly retain the right to practice the concept in the license agreement. In granting a nonexclusive license, the designer remains free to grant licenses to others.

In addition to granting a license on an exclusive or nonexclusive basis, the designer can also restrict the license to a specific activity known as "field of use" or to a specific territory. For example, the designer can grant the right to use the concept, without the right to make or sell it. The field of use can be further divided, such as between consumer and industrial uses. The designer can also grant licenses for a particular geographic territory, such as by city, state, region, or country.

The designer will also need to consider the duration of the license in view of his or her licensing goals. Is the designer seeking short term or long-term licensing objectives? The license can generally be for any period of time through the term of the intellectual property covering the concept. If the license agreement does not specify the duration, the license is usually good for the term of the intellectual property protection. From a practical stand point, the duration of the license is governed by the period that the intellectual property is in effect, the licensee's plans for commercialization of the concept, and the relative obsolescence of the concept. Keep in mind that there are restrictions on the period for which royalties can be charged under a patent license. It is generally a violation of the U.S. antitrust laws to require payment of royalties after a patent has expired, unless the license includes other intellectual property rights such as trade secret know-how or trademarks.

Perhaps the most important factor to the designer in licensing his or her concept is the amount that the licensee will pay for the license. Licenses customarily provide for the payment of running royalties based on a percentage of the sales of product containing the concept; although, in many cases upfront fees and milestone payments are required. A common issue facing licensors in evaluating a license is the royalty rate that the licensee should pay for the license. Each license has to be evaluated on a case-by-case basis because there is no magic royalty rate for all situations.

The royalty rate is determined by a number of factors, such as: the custom of the industry in which the licensed concept will be used; whether the license is exclusive or non-exclusive; whether the license includes a variety of intellectual property rights such as patents, trademarks, trade secrets, and technical assistance; whether the license restricts the scope of the licensee's use of the concept; and whether the size of the market and its pricing is sufficient to support the royalty charged for the license. Typically, royalty rates for general concepts which require significant refinements by the licensee range from about 1 to 5 percent, whereas royalties for finalized and proven concepts usually run from about 5 to 15 percent.

The designer should also consider the following issues in structuring the royalty arrangement for licensing his or her concept:

❏ The time and frequency of royalty payments

❏ The requirement that the licensee keep accurate and complete accounting records

❏ The right of the licensor to inspect the licensee's records

❑ Definition of the net sales or net profits against which the royalties are to be calculated

Many other terms and conditions may be included in the license as deemed necessary by the designer and the licensee, but the license must avoid provisions which violate the antitrust laws or which constitute patent misuse. It has already been explained that a patent license cannot provide for royalty payments after the patent has expired. The license must also avoid illegal price-fixing between the designer and licensee. Tying arrangements should also be avoided, such as where the licensee is required to purchase unprotected products from the licensor to maintain a patent or other intellectual property license.

Selecting the right licensee is essential for the designer to be successful in licensing his or her concept. The designer would want to grant licenses only to reputable companies which are seriously committed to commercializing and marketing the concept. Being viewed as a leader in a particular market or industry, possessing an entrepreneurial spirit, being receptive to new ideas, and having the demonstrated ability to successfully launch new products are attributes which should be considered in the selection of a licensee.

Being able to negotiate effectively is an important part of the designer's success in licensing and profiting from his or her concept. Even for the most successful designers, it may be difficult, because of lack of experience in the art of negotiation and legal and business acumen, to handle a licensing transaction. While licensing is presented very basically in this chapter, it can however in practice be very difficult and complex. To the extent that the licensing deal takes on complex issues different from what is outlined in this chapter, it is advisable that the designer hire an attorney to assist in the structuring and review of any license transaction.

Filling in the Form

In the preamble, begin the preparation by filling in the date of the agreement and the names and addresses of the designer and the licensee.

In paragraph 1, indicate whether the designer is granting the licensee an exclusive or a non-exclusive license. Fill in the title, if applicable, and the description of the design or concept that is being licensed. Mark the appropriate box(es) to indicate if the licensee has the right to make, use, import, or sell the licensed design. Also include the geographical area and the period of time where the license is to be effective.

In paragraph 3, fill in the amount of any nonrefundable advance that the licensee will pay the designer upon signing of the license agreement. Additionally, specify the royalty rate as a percent of net sales of the licensed product that the licensee will pay to the designer. Where the licensed product will not be sold, but will be used or otherwise transferred by the licensee, indicate the percentage increase over the licensee's cost of acquisition or manufacture that will be used in calculating net sales.

Mark the appropriate box in paragraph 4 to indicate whether the licensee will make reports to the designer on a monthly, quarterly, or yearly basis.

In paragraph 6, specify the number of samples of the licensed product that the licensee will give to the designer at no cost. Further, indicate the percentage increase above the licensee's manufacturing costs at which the designer can purchase additional samples.

In paragraph 15, put in the name of the state whose laws will govern the license agreement. Also include the name of the state where the licensee consents to the jurisdiction of the courts.

Include in paragraph 16 the addresses of the designer and licensee where notices, payments, and statements shall be sent. Include in paragraph 17 any additional terms and conditions which are a part of the license agreement.

At the end of the license agreement, fill in the names of the designer and the licensee. Have the agreement signed by authorized signatories for the designer and licensee, and include the signatories' titles.

Negotiation Checklist

❏ Decide what the effective date of the licensing agreement will be, and make sure that the parties are clearly identified by names and location.

❏ Determine whether the licensee is being granted a nonexclusive or an exclusive license and structure the rest of the agreement accordingly. The designer should negotiate for a nonexclusive license if he or she wants freedom to license others.

❏ State the title, if any, of the design concept and include a clear and complete description (including any patent, trademark, and copyright registration numbers).

❏ Decide whether the license includes the broad rights to make, use, import, sell, copy, and distribute licensed products, or if these rights are restricted to certain activities after taking into consideration the designer's overall plans regarding the marketing or licensing of the design.

❏ Decide what the geographical area of the license will be after understanding the potential markets for the licensed products.

❏ Decide on the period of time during which the license will be effective.

❏ Reserve all patents, copyrights, trademarks and all other intellectual property rights in the design to the designer.

❏ Require that credit and copyright notice (if the design is copyrightable) appear in the designer's name on all licensed products. Require that the patent numbers (if the design is patented) appear on the licensed products.

❏ Require the licensee to pay an advance to the designer, which will be applied against the royalties due.

❏ Indicate that any advance paid the licensee is nonrefundable.

❏ Specify a royalty based as a percentage of net sales of the licensed product.

❏ Define net sales to mean invoice price , after deduction of regular trade and quantity discounts, freight, insurance, rebates, returns, allowances, sales and use taxes, and agents' commissions where separately identified in the invoice.

❏ Provide that net sales for licensed products which are not sold, but are otherwise used or otherwise transferred, means the net selling price at which products of similar kind, quality, and quantity are being offered for sale by licensee.

❏ Provide that where licensed products are not being offered for sale by the licensee, net sales for licensed products that are otherwise used or transferred shall be the licensee's cost of acquisition or manufacture, determined by U.S. generally accepted accounting standards, increased by a specified percent.

❏ Require that royalties shall be deemed to accrue when the licensed products are sold, shipped, invoiced, used or otherwise transferred, whichever first occurs.

❏ Require that the licensee provide written royalty reports to the designer on a regular basis; preferably monthly or quarterly, and yearly only if demanded by the licensee.

❏ Require that the licensee include any royalty payments which are due along with the regular royalty reports.

❏ Specify the information to be contained in the royalty report, such as the number, description, and aggregate net sales of licensed products sold, used, or transferred during the reporting period.

❏ Make sure that the designer has the right to terminate the license after thirty days notice if the licensee does not make the required payments.

❏ Provide that all rights granted under the license agreement shall immediately revert to the designer if the agreement is terminated.

❏ Require that the licensee maintain adequate records of the licensed products sold, used, or transferred.

❏ Give the designer the right to inspect the licensee's books and records after the designer has given prior written notice.

❏ Negotiate for the designer to receive a certain number of licensed products at no cost.

❏ Give the designer the right to purchase additional samples of the licensed products at the licensee's manufacturing costs increased by a specified reasonable percentage.

❏ Provide that the designer shall have the right to approve the quality of any reproductions of the design and that the designer's approval will not be unreasonably withheld.

❏ Obligate the licensee to use its best efforts to promote, distribute, and sell the licensed products.

❏ Provide that the licensee's use of the design shall inure to the designer's benefit if the licensee acquires any trademarks or other rights in and to the design.

❏ Require the licensee to assign and transfer any trademark or other rights in the design to the designer upon termination of the license agreement.

❏ Make it clear that all rights not specifically transferred by the license agreement are reserved to the designer.

❏ Require the licensee to indemnify the designer against any costs arising out of the use of the design for the licensed products.

❏ Limit the designer's liabilities by providing no warranties or representations (including infringement of any third parties' rights, and assuming no responsibilities with respect to any product containing the design.

❏ Restrict assignment of the license by providing that neither the designer nor the licensee can assign the license, except that the designer should be free to assign rights to any payments due under the license agreement.

❏ Provide that nothing in the license agreement shall constitute a joint venture or any similar relationship between the designer and licensee.

❏ Choose the state where the designer is located or most familiar as the place where the agreement will be interpreted and governed.

❏ Recite that the agreement is binding on the client and designer, as well as other parties who may stand in their place, such as heirs, successors, assigns, and personal representatives.

❏ Include in the agreement any additional terms and conditions (such as minimum royalties, insurance, etc.) that are believed to be necessary or desirable.

❏ Make sure that the agreement reflects the entire understanding of the parties, and that modifications can only be made in writing and signed by both parties.

❏ Make sure that the agreement is properly executed by authorized signatories for both parties.

❏ Compare paragraphs 13–18 to the standard provisions in the introduction.

Licensing Agreement for Industrial Designs

AGREEMENT, dated the _____ day of _____, 20 _____, between _____ (hereinafter referred to as the "Designer"), located at _____ and _____ (hereinafter referred to as the "Licensee"), located at _____, with respect to the use of a certain design created by the Designer (hereinafter referred to as the "Design") for manufactured products (hereinafter referred to as the "Licensed Products").

WHEREAS, the Designer is a professional designer of good standing; and

WHEREAS, the Designer has created the Design which the Designer wishes to license for purposes of manufacture and sale; and

WHEREAS, the Licensee wishes to use the Design to create a certain product or products for manufacture and sale; and

WHEREAS, both parties want to achieve the best possible quality to generate maximum sales;

NOW, THEREFORE, in consideration of the foregoing premises and the mutual covenants hereinafter set forth and other valuable consideration, the parties hereto agree as follows:

1.   **Grant of Rights to the Design.** The Designer grants to the Licensee the ❏ exclusive ❏ nonexclusive license to the Design, titled _____, and described as _____, and which was created and is owned by the Designer; wherein said license consists of the right by the Licensee to ❏ make, ❏ use, ❏ import, ❏ sell, ❏ copy, ❏ distribute, and ❏ other (specify: _____) by the Licensee in the following geographical area _____ and for the following period of time _____.

2.   **Ownership of Intellectual Property.** The Designer shall retain all intellectual property in and to the Design, including, but not limited to patents, copyrights, trademarks, and know-how. The Licensee shall identify the Designer as the creator of the Design on the Licensed Products and shall, if the Design is copyrightable, reproduce thereon a copyright notice for the Designer, which shall include the word "Copyright" or the symbol for copyright, the Designer's name, and the year of first publication. If the Design is patented, the Licensee shall include the applicable patent numbers on the Licensed Products.

3.   **Advance and Royalties.** The Licensee agrees to pay the Designer a nonrefundable advance in the amount of $_____ upon signing this Agreement, which advance shall be applied against royalties due to the Designer hereunder. The Licensee further agrees to pay the Designer a royalty of _____ percent of the Net Sales of the Licensed Products. "Net Sales" for Licensed Products sold means invoice price, after deduction of regular trade and quantity discounts, freight, insurance, rebates, returns, allowances, sales and use taxes, and agents' commissions where separately identified in the invoice. "Net Sales" for Licensed Products which are not sold, but are otherwise used or otherwise transferred, means the net selling price at which products of similar kind, quality, and quantity are being offered for sale by the Licensee. Where such products are not being offered for sale by the Licensee, "Net Sales" for Licensed Products which are otherwise used or transferred shall be the Licensee's cost of acquisition or manufacture, determined by U.S. generally accepted accounting standards, increased by _____ percent. Royalties shall be deemed to accrue when the Licensed Products are sold, shipped, invoiced, used, or otherwise transferred, whichever first occurs.

4. Reports and Payments. The Licensee shall make written reports to the Designer within thirty (30) days after the end of each calendar ❑ month ❑ quarter ❑ year stating the number, description and aggregate Net Sales of Licensed Products sold, used or otherwise transferred during that period. Along with each report, Licensee shall pay to the Designer any and all royalties due and payable to the Designer for the Licensed Products included in the report. The Designer shall have the right to terminate this Agreement upon thirty (30) days notice if the Licensee fails to make any payment required of it and does not cure this default within said thirty (30) days, whereupon all rights granted herein shall revert immediately to the Designer.

5. Inspection of Books and Records. The Licensee shall keep adequate and accurate records showing the Licensed Products sold, used or otherwise transferred. Upon prior written notice, Designer shall have the right to inspect Licensee's books and records concerning the Licensed Products.

6. Samples. The Licensee shall give at no cost to the Designer _____ samples of the Licensed Products for the Designer's use. The Designer shall have the right to purchase additional samples of the Licensed Products at the Licensee's manufacturing cost increased by _____ percent.

7. Quality of Reproductions. The Designer shall have the right to approve the quality of the reproduction of the Design on the Licensed Products, and the Designer agrees not to withhold approval unreasonably.

8. Promotion. The Licensee shall use its best efforts to promote, distribute, and sell the Licensed Products.

9. Trademarks and Other Rights. The Licensee's use of the Design shall inure to the benefit of the Designer if the Licensee acquires any trademarks, trade rights, equities, titles, or other rights in and to the Design whether by operation of law, usage, or otherwise during the term of this Agreement or any extension thereof. Upon the expiration of this Agreement or any extension thereof or sooner termination, the Licensee shall assign and transfer the said trademarks, trade rights, equities, titles, or other rights to the Designer without any additional consideration.

10. Reservation of Rights. All rights not specifically transferred by this Agreement are reserved to the Designer.

11. Indemnification. The Licensee shall hold the Designer harmless from and against any loss, expense, or damage occasioned by any claim, demand, suit, or recovery against the Designer arising out of the use of the Design for the Licensed Products.

12. Limitations. Nothing in this agreement shall be construed as a warranty or representation by the Designer that the Design or Licensed Products is or will be free from the infringement of any proprietary rights of third persons. The Designer makes no representations, extends no warranties of any kind, and assumes no responsibilities with respect to use, sale, or other disposition of any product containing the Design.

13. Assignment. Neither party shall assign rights or obligations under this Agreement, except that the Designer may assign the right to receive money due hereunder.

14. Nature of Contract. Nothing herein shall be construed to constitute the parties hereto as joint ventures, nor shall any similar relationship be deemed to exist between them.

15. **Governing Law.** This Agreement shall be construed in accordance with the laws of the State of _____; the Licensee consents to the jurisdiction of the courts of the State of _____.

16. **Addresses.** All notices, demands, payments, royalty payments, and statements shall be sent to the Designer at the following address_____ and to the Licensee at _____.

17. **Additional Terms and Conditions:**

_____

_____

_____

_____

_____

18. **Assigns/Modifications.** This Agreement shall be binding upon the parties hereto, their heirs, successors, assigns, and personal representatives. This Agreement constitutes the entire agreement between the parties hereto and shall not be modified, amended, or changed in any way except by a written agreement signed by both parties hereto.

IN WITNESS WHEREOF, the parties have signed this Agreement as of the date first set forth above.

Designer _____   Licensee _____

By _____   By _____
    Authorized Signatory, Title   Authorized Signatory, Title

Designer's Lecture Contract

Many industrial designers find lecturing to be both a source of income and a rewarding opportunity to express their feelings about their work and their profession. High schools, colleges, conferences, professional societies, and other institutions often invite designers to lecture. Slides of their designs may be used during these lectures, and in some cases, an exhibition may be mounted during the designer's visit.

A contract ensures that everything goes smoothly. For example, who should pay for slides that the designer has to make for that particular lecture? Who will pay for transportation to and from the lecture? Who will supply materials for a demonstration of technique? Will the designer have to give one lecture in a day or, as the institution might prefer, many more? Will the designer have to review portfolios of students? Resolving these kinds of questions, as well as the amount of and time in which to pay the fee, will make any lecture a more rewarding experience.

Filling in the Form

In the preamble, give the date and the names and addresses of the parties. In paragraph 1, give the dates when the designer will lecture, the nature and extent of the services the designer will perform, and the form in which the designer is to bring examples of his or her work. In paragraph 2, specify the fee to be paid to the designer and when it will be paid during the designer's visit.

In paragraph 3, give the amounts of expenses to be paid (or state that none or all of these expenses are to be paid); specify which expenses other than travel, food, and lodging are covered; and show what will be provided by the sponsor (such as food or lodging). In paragraph 10, indicate which state's law will govern the contract. Then have both parties sign the con-

tract. On the schedule of designs, list the works (such as computer-assisted renderings, photographs, prototypes, and the like) to be used for illustration and their insurance value.

Negotiation Checklist

❏ How long will the designer be required to stay at the sponsoring institution in order to perform the required services? (Paragraph 1)

❏ What are the nature and extent of the services the designer will be required to perform? (Paragraph 1)

❏ What slides, original designs, or other materials must the designer bring? (Paragraph 1)

❏ Specify the work facilities that the sponsor will provide the designer.

❏ Specify the fee to be paid to the designer. (Paragraph 2)

❏ Give the time to pay the fee. (Paragraph 2)

❏ Require part of the fee be paid in advance.

❏ Specify the expenses that will be paid by the sponsor, including the time for payment of these expenses. (Paragraph 3)

❏ Indicate what the sponsor may provide in place of paying expenses, such as giving lodging, meals, or a car. (Paragraph 3)

❏ If illness prevents the designer from coming to lecture, state that an effort will be made to find another date. (Paragraph 4)

❏ If the sponsor must cancel for a reason beyond its control, indicate that the expenses incurred by the designer must be paid and there will be an attempt to reschedule. (Paragraph 4)

❏ If the sponsor cancels within forty-eight hours of the time the designer is to arrive, consider requiring that the full fee as well as expenses be paid.

❏ Provide for the payment of interest on late payments by the sponsor. (Paragraph 5)

❏ Retain for the designer all rights, including copyrights, in any recordings of any kind that may be made of the designer's visit. (Paragraph 6)

❏ If the sponsor wishes to use a recording of the designer's visit, such as a video, require that the sponsor obtain the designer's written permission and that, if appropriate, a fee be negotiated for this use. (Paragraph 6)

❏ Provide that the sponsor is strictly responsible for loss or damage to any designs from the time they leave the designer's studio until they are returned there. (Paragraph 7)

❏ Require the sponsor to insure the designs, and specify insurance values. (Paragraph 7)

❏ Consider which risks may be excluded from the insurance coverage.

❏ Consider whether the designer should be the named beneficiary of the insurance coverage for his or her works.

❏ State who will pay the cost of packing and shipping the works to and from the sponsor. (Paragraph 8)

❏ State who will take the responsibility to pack and ship the works to and from the sponsor.

❏ Compare the standard provisions in the introductory pages with paragraphs 9 and 10.

Designer's Lecture Contract

AGREEMENT, dated the _____ day of _____, 20 _____, between _____ (hereinafter referred to as the "Designer"), located at _____ and _____ (hereinafter referred to as the "Sponsor"), located at _____.

WHEREAS, the Sponsor is familiar with and admires the work of the Designer; and

WHEREAS, the Sponsor wishes the Designer to visit the Sponsor to enhance the opportunities for its students to have contact with working professional designer; and

WHEREAS, the Designer wishes to lecture with respect to his or her work and perform such other services as this contract may call for;

NOW, THEREFORE, in consideration of the foregoing premises and the mutual covenants hereinafter set forth and other valuable considerations, the parties hereto agree as follows:

1.   Designer to Lecture. The Designer hereby agrees to come to the Sponsor on the following date(s): _____ and perform the following services: _____.

The Designer shall use best efforts to make his or her services as productive as possible to the Sponsor. The Designer further agrees to bring examples of his or her own work in the form of _____ _____.

2.   Payment. The Sponsor agrees to pay as full compensation for the Designer's services rendered under Paragraph 1 the sum of $_____. This sum shall be payable to the Designer on completion of the _____ day of the Designer's residence with the Sponsor.

3.   Expenses. In addition to the payments provided under Paragraph 2, the Sponsor agrees to reimburse the Designer for the following expenses:

(A) Travel expenses in the amount of $_____

(B) Food and lodging expenses in the amount of $_____

(C) Other expenses listed here: _____ in the amount of $_____

The reimbursement for travel expenses shall be made fourteen (14) days prior to the earliest date specified in Paragraph 1. The reimbursement for food, lodging, and other expenses shall be made at the date of payment specified in Paragraph 2, unless a contrary date is specified here _____.

In addition, the Sponsor shall provide the Designer with the following:

(A) Tickets for travel, rental car, or other modes of transportation as follows

_____
_____

(B) Food and lodging as follows

_____

_____

(C) Other hospitality as follows

_____

_____

4.   Inability to Perform. If the Designer is unable to appear on the dates scheduled in Paragraph 1 due to illness, the Sponsor shall have no obligation to make any payments under Paragraphs 2 and 3, but shall attempt to reschedule the Designer's appearance at a mutually acceptable future date. If the Sponsor is prevented from having the Designer appear by acts of God, hurricane, flood, governmental order, or other cause beyond its control, the Sponsor shall be responsible only for the payment of such expenses under Paragraph 3 as the Designer shall have actually incurred. The Sponsor agrees in such a case to attempt to reschedule the Designer's appearance at a mutually acceptable future date.

5.   Late Payment. The Sponsor agrees that, in the event it is late in making payment of amounts due to the Designer under Paragraphs 2,3, or 8, it will pay as additional liquidated damages _____ percent in interest on the amounts it is owing to the Designer, said interest to run from the date stipulated for payment in Paragraphs 2, 3, or 8 until such time as payment is made.

6.   Copyrights and Recordings. Both parties agree that the Designer shall retain all rights, including copyrights, in relation to recordings of any kind made of the appearance or any works shown in the course thereof. The term "recording" as used herein shall include any recording made by electronic transcription, tape recording, wire recording, film, videotape, or other similar or dissimilar methods of recording, whether now known or hereinafter developed. No use of any such recording shall be made by the Sponsor without the written consent of the Designer and, if stipulated therein, additional compensation for such use.

7.   Insurance and Loss or Damage. The Sponsor agrees that it shall provide wall-to-wall insurance for the works listed on the Schedule of Designs for the values specified therein. The Sponsor agrees that it shall be fully responsible and have strict liability for any loss or damage to the designs from the time said designs leaves the Designer's residence or studio until such time as they are returned there.

8.   Packing and Shipping. The Sponsor agrees that it shall fully bear any costs of packing and shipping necessary to deliver the works specified in Paragraph 7 to the Sponsor and return them to the Designer's residence or studio.

9.   Modification. This contract contains the full understanding between the parties hereto and may only be modified in a written instrument signed by both parties.

10.   Governing Law. This contract shall be governed by the laws of the State of _____.

IN WITNESS WHEREOF, the parties hereto have signed this Agreement as of the date first set forth above.

Designer _____   Sponsor _____
                Company Name

By _____
        Authorized Signatory, Title

Schedule of Designs

	Title	Description	Size	Value
1.				
2.				
3.				
4.				

Copyright Application Form and Short Form Application

The industrial designer may be able to protect proprietary designs via copyright. However, copyright protection is limited and applies only to the artistic and literary features of the design. Unlike a utility patent that covers the concept itself, a copyright covers only the designer's particular form of expression of the concept. Copyright does not prevent others from using the utilitarian features of the design or information revealed by the design. Thus, although a design concept may have copyright protection, anyone is free to use the functional features of the design concept, and may create his or her own expression of the same concepts as long as he or she does not copy the designer's artistic or literary features.

A copyright gives the designer the sole right to control the reproduction and distribution of the designer's copyrighted work. The copyright owner has the right to prevent others from reproducing the work without the owner's permission.

The industrial designer should always review his or her creative works to see if they can be protected by copyright. In reference to industrial designs, the federal copyright law permits copyright protection for original literary, pictorial, graphic, and sculptural works. Literary works are works, other than audiovisual works, that are expressed in words, numbers, or other verbal or numerical symbols, such as books, magazines, manuscripts, directories, catalogs, cards, instructional manuals, and computer programs. Pictorial, graphic, and sculptural works include two- and three-dimensional works of art, photographs, paintings, sketches, prints, maps, charts, designs, technical drawings, diagrams, and models.

Some common areas where copyrights and patents overlap are in shapes and designs of containers, devices, apparatus, toys, and games. For example, an artistic bottle shape can be protected by a copyright and a design patent.

A board game, for example, can be covered by a utility patent; the ornamental game board covered by a design patent and a copyright; and the rules, instructions, and packaging covered by copyright. Seeing how copyrights and patents interface, the designer should always consider them in protecting his or her designs.

Copyright protection is only available for original works of authorship. The originality requirement means that the work must be the independent creation of the person seeking copyright protection. The test for originality is met if the work owes its origin to the designer and is independently created and not copied from other works. The mere fact that someone else has created something similar will not prevent the designer from obtaining copyright protection on his or her independent creation.

Copyright protection does not extend to names, titles, and slogans. Where appropriate, these can be protected by trademark.

To qualify for federal copyright protection, the designer's work must be fixed in a tangible form. This means that the work must be in a physical form such that it can be reproduced or otherwise communicated. Thus a mere idea, or other abstract and imaginary creations cannot be copyrighted.

A copyright arises automatically when the work is created and registration is not required for the copyright to come into existence. However, the designer's rights to enforce his or her copyright are maximized if the copyright is registered with the U.S. Copyright Office.

You can register your copyright by filing an application with the U.S. Copyright Office. This involves filing the appropriate application form along with the required fee ($30 at the time of this printing) and deposit of your work.

Anyone who violates any of the exclusive rights of the copyright owner to reproduce, distribute, display or make derivatives of the copyrighted work is an infringer. The copyright owner may recover from an infringer statutory damages or actual damages plus any profits of the infringer. The copyright owner may also recover court costs and attorneys fees from the infringer if allowed by the court. The copyright owner may also obtain an injunction to prevent and restrain infringement of the copyright. Registration of the copyrighted work prior to the infringement is necessary to be eligible for statutory damages, court costs and attorney fees. Copyright registration must also be obtained before any copyright infringement action can be brought. Copyright infringement actions must be brought within three years after the infringement takes place or should have been discovered.

Fair use of copyrighted work is permitted and does not amount to copyright infringement. Fair use for purposes such as criticism, comment, news reporting, teaching, or research is not an infringement of copyright.

The copyright law requires a specific notice to enable the owner to obtain the strongest copyright protection for published works. If the designer plans to publish or publicly distribute his or her work, he should place a copyright notice on all such copies. Use of the copyright notice does not require the permission of the Copyright Office, and the notice can be used even if the work has not been registered with the U.S. Copyright Office. Since March 1989, the copyright owner will not lose his or her copyright if he fails to put the copyright notice on his or her works. The benefit of the copyright notice is that infringers cannot ask the court to lessen the amount of damages because they were unaware of the owner's copyright.

The copyright notice consists of three parts: 1) the copyright symbol © or the word "copyright", 2) the year of the first publication of the work, and 3) the name of the copyright owner. For example, the proper copyright notice could follow this form: "Copyright 2004 ABC Designs, Inc." The notice should be affixed to the work in such a manner and location as to give reasonable notice of the claim of copyright. Note that a letter C in parentheses is not a valid substitute for the copyright symbol.

A copyright registration is effective as of the date that the U.S. Copyright Office receives the application, fee, and deposit materials in an acceptable form, regardless of how long it takes to receive the certificate of registration. To know whether the Copyright Office received the registration materials, the designer can send them by registered or certified mail with a return receipt requested from the post office.

Filling in Form VA and Short Form VA

The designer can register his or her copyright for visual arts by filing an application with the U.S. Copyright Office. This involves filing the application Form VA or Short Form VA along with the required fee and deposit of the work.

Detailed instructions for completing the copyright application are given at the end of this chapter along with a copy of Form VA or Short Form VA from the U.S. Copyright Office. The instructions are also useful for Form PA (performing art), Form SR (sound recording), and Form TX, which is used to register textual and nondramatic literary works, such as operating manuals, product instructions, advertisements, and computer programs.

The proper deposit for the copyright registration is usually two complete copies of published works. For unpublished works, only one complete copy is required for registration. If the work is unique or a limited edition, the designer can submit photos, photocopies or other identifying materials instead of an original copy.

The Copyright Office will not accept the copyright application unless it contains the date and the handwritten signature of the author or other copyright claimant, or of the owner of exclusive rights, or of the duly authorized agent of the author, claimant, or owner of exclusive rights.

For additional information on copyrights refer to *The Copyright Guide: A Friendly Handbook for Protecting and Profiting from Copyrights*, Third Edition, by Lee Wilson (Allworth Press, 2003). Additionally, the Copyright Office has an information telephone (202-707-3000) and also makes available a free Copyright Information Kit, which contains copies of Form VA and other Copyright Office circulars. Copyright forms and circulars can also be requested by contacting the hotline number (202-707-9100) or by downloading these from the Copyright Office Web site (*www.copyright.gov*).

Checklist for Registering a Copyright

❏ Fill out appropriate application form, and make sure that all applicable spaces have been filled in.

❏ Use Form VA or Short Form VA for works of the visual arts such as pictorial, graphic, and sculptural works.

❏ Use Form TX or Short Form TX for non-dramatic literary and textual works, such as books, directories, advertising copy, and compilations of information.

❏ Sign the form in the space provided for certification.

❏ Enclose with the copyright application a check or money order for the $30 application fee payable to Register of Copyrights.

❏ Mail the completed and signed application, the $30 application fee, and the necessary copies of the work (two copies for published works; one copy for unpublished works) to:

 Register of Copyrights

 Library of Congress

 101 Independence Ave., SE

 Washington, DC 20559–6000

❏ Consult an attorney if the copyright registration involves difficult or complex issues.

Form VA

Detach and read these instructions before completing this form.
Make sure all applicable spaces have been filled in before you return this form.

BASIC INFORMATION

When to Use This Form: Use Form VA for copyright registration of published or unpublished works of the visual arts. This category consists of "pictorial, graphic, or sculptural works," including two-dimensional and three-dimensional works of fine, graphic, and applied art, photographs, prints and art reproductions, maps, globes, charts, technical drawings, diagrams, and models.

What Does Copyright Protect? Copyright in a work of the visual arts protects those pictorial, graphic, or sculptural elements that, either alone or in combination, represent an "original work of authorship." The statute declares: "In no case does copyright protection for an original work of authorship extend to any idea, procedure, process, system, method of operation, concept, principle, or discovery, regardless of the form in which it is described, explained, illustrated, or embodied in such work."

Works of Artistic Craftsmanship and Designs: "Works of artistic craftsmanship" are registrable on Form VA, but the statute makes clear that protection extends to "their form" and not to "their mechanical or utilitarian aspects." The "design of a useful article" is considered copyrightable "only if, and only to the extent that, such design incorporates pictorial, graphic, or sculptural features that can be identified separately from, and are capable of existing independently of, the utilitarian aspects of the article."

Labels and Advertisements: Works prepared for use in connection with the sale or advertisement of goods and services are registrable if they contain "original work of authorship." Use Form VA if the copyrightable material in the work you are registering is mainly pictorial or graphic; use Form TX if it consists mainly of text. **NOTE:** Words and short phrases such as names, titles, and slogans cannot be protected by copyright, and the same is true of standard symbols, emblems, and other commonly used graphic designs that are in the public domain. When used commercially, material of that sort can sometimes be protected under state laws of unfair competition or under the federal trademark laws. For information about trademark registration, write to the Commissioner of Patents and Trademarks, Washington, D.C. 20231.

Architectural Works: Copyright protection extends to the design of buildings created for the use of human beings. Architectural works created on or after December 1, 1990, or that on December 1, 1990, were unconstructed and embodied only in unpublished plans or drawings are eligible. Request Circular 41 for more information. Architectural works and technical drawings cannot be registered on the same application.

Deposit to Accompany Application: An application for copyright registration must be accompanied by a deposit consisting of copies representing the entire work for which registration is to be made.

> **Unpublished Work:** Deposit one complete copy.
>
> **Published Work:** Deposit two complete copies of the best edition.
>
> **Work First Published Outside the United States:** Deposit one complete copy of the first foreign edition.
>
> **Contribution to a Collective Work:** Deposit one complete copy of the best edition of the collective work.

The Copyright Notice: Before March 1, 1989, the use of copyright notice was mandatory on all published works, and any work first published before that date should have carried a notice. For works first published on and after March 1, 1989, use of the copyright notice is optional. For more information about copyright notice, see Circular 3, "Copyright Notice."

For Further Information: To speak to an information specialist, call (202) 707-3000 (TTY: (202) 707-6737). Recorded information is available 24 hours a day. Order forms and other publications from the address in space 9 or call the Forms and Publications Hotline at (202) 707-9100. Most circulars (but not forms) are available via fax. Call (202) 707-2600 from a touchtone phone. Access and download circulars, forms, and other information from the Copyright Office website at *www.copyright.gov*.

LINE-BY-LINE INSTRUCTIONS
Please type or print using black ink. The form is used to produce the certificate.

1 SPACE 1: Title

Title of This Work: Every work submitted for copyright registration must be given a title to identify that particular work. If the copies of the work bear a title (or an identifying phrase that could serve as a title), transcribe that wording *completely* and *exactly* on the application. Indexing of the registration and future identification of the work will depend on the information you give here. For an architectural work that has been constructed, add the date of construction after the title; if unconstructed at this time, add "not yet constructed."

Publication as a Contribution: If the work being registered is a contribution to a periodical, serial, or collection, give the title of the contribution in the "Title of This Work" space. Then, in the line headed "Publication as a Contribution," give information about the collective work in which the contribution appeared.

Nature of This Work: Briefly describe the general nature or character of the pictorial, graphic, or sculptural work being registered for copyright. Examples: "Oil Painting"; "Charcoal Drawing"; "Etching"; "Sculpture"; "Map"; "Photograph"; "Scale Model"; "Lithographic Print"; "Jewelry Design"; "Fabric Design."

Previous or Alternative Titles: Complete this space if there are any additional titles for the work under which someone searching for the registration might be likely to look, or under which a document pertaining to the work might be recorded.

2 SPACE 2: Author(s)

General Instruction: After reading these instructions, decide who are the "authors" of this work for copyright purposes. Then, unless the work is a "collective work," give the requested information about every "author" who contributed any appreciable amount of copyrightable matter to this version of the work. If you need further space, request Continuation Sheets. In the case of a collective work, such as a catalog of paintings or collection of cartoons by various authors, give information about the author of the collective work as a whole.

Name of Author: The fullest form of the author's name should be given. Unless the work was "made for hire," the individual who actually created the work is its "author." In the case of a work made for hire, the statute provides that "the employer or other person for whom the work was prepared is considered the author."

What is a "Work Made for Hire"? A "work made for hire" is defined as: (1) "a work prepared by an employee within the scope of his or her employment"; or (2) "a work specially ordered or commissioned for use as a contribution to a collective work, as a part of a motion picture or other audiovisual work, as a translation, as a supplementary work, as a compilation, as an instructional text, as a test, as answer material for a test, or as an atlas, if the parties expressly agree in a written instrument signed by them that the work shall be considered a work made for hire." If you have checked "Yes" to indicate that the work was "made for hire," you must give the full legal name of the employer (or other person for whom the work was prepared). You may also include the name of the employee along with the name of the employer (for example: "Elster Publishing Co., employer for hire of John Ferguson").

"Anonymous" or "Pseudonymous" Work: An author's contribution to a work is "anonymous" if that author is not identified on the copies or phonorecords of the work. An author's contribution to a work is "pseudonymous" if that author is identified on the copies or phonorecords under a fictitious name. If the work is "anonymous" you may: (1) leave the line blank; or (2) state "anonymous" on the line; or (3) reveal the author's identity. If the work is "pseudonymous" you may: (1) leave the line blank; or (2) give the pseudonym and identify it as such (for example: "Huntley Haverstock, pseudonym"); or (3) reveal the author's name, making clear which is the real name and which is the pseudonym (for example: "Henry Leek, whose pseudonym is Priam Farrel"). However, the citizenship or domicile of the author **must** be given in all cases.

Dates of Birth and Death: If the author is dead, the statute requires that the year of death be included in the application unless the work is anonymous or pseudonymous. The author's birth date is optional but is useful as a form of identification. Leave this space blank if the author's contribution was a "work made for hire."

Author's Nationality or Domicile: Give the country of which the author is a citizen or the country in which the author is domiciled. Nationality or domicile **must** be given in all cases.

Nature of Authorship: Catagories of pictorial, graphic, and sculptural authorship are listed below. Check the box(es) that best describe(s) each author's contribution to the work.

3-Dimensional sculptures: fine art sculptures, toys, dolls, scale models, and sculptural designs applied to useful articles.

2-Dimensional artwork: watercolor and oil paintings; pen and ink drawings; logo illustrations; greeting cards; collages; stencils; patterns; computer graphics; graphics appearing in screen displays; artwork appearing on posters, calendars, games, commercial prints and labels, and packaging, as well as 2-dimensional artwork applied to useful articles, and designs reproduced on textiles, lace, and other fabrics; on wallpaper, carpeting, floor tile, wrapping paper, and clothing.

Reproductions of works of art: reproductions of preexisting artwork made by, for example, lithography, photoengraving, or etching.

Maps: cartographic representations of an area, such as state and county maps, atlases, marine charts, relief maps, and globes.

Photographs: pictorial photographic prints and slides and holograms.

Jewelry designs: 3-dimensional designs applied to rings, pendants, earrings, necklaces, and the like.

Technical drawings: diagrams illustrating scientific or technical information in linear form, such as architectural blueprints or mechanical drawings.

Text: textual material that accompanies pictorial, graphic, or sculptural works, such as comic strips, greeting cards, games rules, commercial prints or labels, and maps.

Architectural works: designs of buildings, including the overall form as well as the arrangement and composition of spaces and elements of the design.

NOTE: Any registration for the underlying architectural plans must be applied for on a separate Form VA, checking the box "Technical drawing."

SPACE 3: Creation and Publication

General Instructions: Do not confuse "creation" with "publication." Every application for copyright registration must state "the year in which creation of the work was completed." Give the date and nation of first publication only if the work has been published.

Creation: Under the statute, a work is "created" when it is fixed in a copy or phonorecord for the first time. Where a work has been prepared over a period of time, the part of the work existing in fixed form on a particular date constitutes the created work on that date. The date you give here should be the year in which the author completed the particular version for which registration is now being sought, even if other versions exist or if further changes or additions are planned.

Publication: The statute defines "publication" as "the distribution of copies or phonorecords of a work to the public by sale or other transfer of ownership, or by rental, lease, or lending"; a work is also "published" if there has been an "offering to distribute copies or phonorecords to a group of persons for purposes of further distribution, public performance, or public display." Give the full date (month, day, year) when, and the country where, publication first occurred. If first publication took place simultaneously in the United States and other countries, it is sufficient to state "U.S.A."

SPACE 4: Claimant(s)

Name(s) and Address(es) of Copyright Claimant(s): Give the name(s) and address(es) of the copyright claimant(s) in this work even if the claimant is the same as the author. Copyright in a work belongs initially to the author of the work (including, in the case of a work made for hire, the employer or other person for whom the work was prepared). The copyright claimant is either the author of the work or a person or organization to whom the copyright initially belonging to the author has been transferred.

Transfer: The statute provides that, if the copyright claimant is not the author, the application for registration must contain "a brief statement of how the claimant obtained ownership of the copyright." If any copyright claimant named in space 4 is not an author named in space 2, give a brief statement explaining how the claimant(s) obtained ownership of the copyright. Examples: "By written contract"; "Transfer of all rights by author"; "Assignment"; "By will." Do not attach transfer documents or other attachments or riders.

SPACE 5: Previous Registration

General Instructions: The questions in space 5 are intended to find out whether an earlier registration has been made for this work and, if so, whether there is any basis for a new registration. As a rule, only one basic

copyright registration can be made for the same version of a particular work.

Same Version: If this version is substantially the same as the work covered by a previous registration, a second registration is not generally possible unless: (1) the work has been registered in unpublished form and a second registration is now being sought to cover this first published edition; or (2) someone other than the author is identified as a copyright claimant in the earlier registration, and the author is now seeking registration in his or her own name. If either of these two exceptions applies, check the appropriate box and give the earlier registration number and date. Otherwise, do not submit Form VA; instead, write the Copyright Office for information about supplementary registration or recordation of transfers of copyright ownership.

Changed Version: If the work has been changed and you are now seeking registration to cover the additions or revisions, check the last box in space 5, give the earlier registration number and date, and complete both parts of space 6 in accordance with the instruction below.

Previous Registration Number and Date: If more than one previous registration has been made for the work, give the number and date of the latest registration.

SPACE 6: Derivative Work or Compilation

General Instructions: Complete space 6 if this work is a "changed version," "compilation," or "derivative work," and if it incorporates one or more earlier works that have already been published or registered for copyright, or that have fallen into the public domain. A "compilation" is defined as "a work formed by the collection and assembling of preexisting materials or of data that are selected, coordinated, or arranged in such a way that the resulting work as a whole constitutes an original work of authorship." A "derivative work" is "a work based on one or more preexisting works." Examples of derivative works include reproductions of works of art, sculptures based on drawings, lithographs based on paintings, maps based on previously published sources, or "any other form in which a work may be recast, transformed, or adapted." Derivative works also include works "consisting of editorial revisions, annotations, or other modifications" if these changes, as a whole, represent an original work of authorship.

Preexisting Material (space 6a): Complete this space **and** space 6b for derivative works. In this space identify the preexisting work that has been recast, transformed, or adapted. Examples of preexisting material might be "Grunewald Altarpiece" or "19th century quilt design." Do not complete this space for compilations.

Material Added to This Work (space 6b): Give a brief, general statement of the **additional** new material covered by the copyright claim for which registration is sought. In the case of a derivative work, identify this new material. Examples: "Adaptation of design and additional artistic work"; "Reproduction of painting by photolithography"; "Additional cartographic material"; "Compilation of photographs." If the work is a compilation, give a brief, general statement describing both the material that has been compiled **and** the compilation itself. Example: "Compilation of 19th century political cartoons."

SPACE 7, 8, 9: Fee, Correspondence, Certification, Return Address

Deposit Account: If you maintain a Deposit Account in the Copyright Office, identify it in space 7a. Otherwise, leave the space blank and send the fee of $30 with your application and deposit.

Correspondence (space 7b): This space should contain the name, address, area code, telephone number, email address, and fax number (if available) of the person to be consulted if correspondence about this application becomes necessary.

Certification (space 8): The application cannot be accepted unless it bears the date and the **handwritten signature** of the author or other copyright claimant, or of the owner of exclusive right(s), or of the duly authorized agent of the author, claimant, or owner of exclusive right(s).

Address for Return of Certificate (space 9): The address box must be completed legibly since the certificate will be returned in a window envelope.

PRIVACY ACT ADVISORY STATEMENT Required by the Privacy Act of 1974 (P.L. 93 - 579)
The authority for requesting this information is title 17, U.S.C., secs. 409 and 410. Furnishing the requested information is voluntary. But if the information is not furnished, it may be necessary to delay or refuse registration and you may not be entitled to certain relief, remedies, and benefits provided in chapters 4 and 5 of title 17, U.S.C.
The principal uses of the requested information are the establishment and maintenance of a public record and the examination of the application for compliance with the registration requirements of the copyright code.
Other routine uses include public inspection and copying, preparation of public indexes, preparation of public catalogs of copyright registrations, and preparation of search reports upon request.
NOTE: No other advisory statement will be given in connection with this application. Please keep this statement and refer to it if we communicate with you regarding this application.

Copyright Office fees are subject to change. For current fees, check the Copyright Office website at *www.copyright.gov*, write the Copyright Office, or call (202) 707-3000.

Ⓒ Form VA
For a Work of the Visual Arts
UNITED STATES COPYRIGHT OFFICE

REGISTRATION NUMBER

VA   VAU

EFFECTIVE DATE OF REGISTRATION

Month   Day   Year

DO NOT WRITE ABOVE THIS LINE. IF YOU NEED MORE SPACE, USE A SEPARATE CONTINUATION SHEET.

1

Title of This Work ▼

NATURE OF THIS WORK ▼ See instructions

Previous or Alternative Titles ▼

Publication as a Contribution If this work was published as a contribution to a periodical, serial, or collection, give information about the collective work in which the contribution appeared. **Title of Collective Work ▼**

If published in a periodical or serial give: **Volume ▼**   **Number ▼**   **Issue Date ▼**   **On Pages ▼**

2

a

NAME OF AUTHOR ▼

DATES OF BIRTH AND DEATH
Year Born ▼   Year Died ▼

NOTE

Under the law, the "author" of a **"work made for hire"** is generally the employer, not the employee (see instructions). For any part of this work that was "made for hire" check "Yes" in the space provided, give the employer (or other person for whom the work was prepared) as "Author" of that part, and leave the space for dates of birth and death blank.

Was this contribution to the work a "work made for hire"?
☐ Yes
☐ No

Author's Nationality or Domicile
Name of Country
OR { Citizen of _____
Domiciled in _____

Was This Author's Contribution to the Work
Anonymous?   ☐ Yes ☐ No
Pseudonymous? ☐ Yes ☐ No

If the answer to either of these questions is "Yes," see detailed instructions.

Nature of Authorship Check appropriate box(es). **See instructions**
☐ 3-Dimensional sculpture   ☐ Map   ☐ Technical drawing
☐ 2-Dimensional artwork   ☐ Photograph   ☐ Text
☐ Reproduction of work of art ☐ Jewelry design ☐ Architectural work

b

Name of Author ▼

Dates of Birth and Death
Year Born ▼   Year Died ▼

Was this contribution to the work a "work made for hire"?
☐ Yes
☐ No

Author's Nationality or Domicile
Name of Country
OR { Citizen of _____
Domiciled in _____

Was This Author's Contribution to the Work
Anonymous?   ☐ Yes ☐ No
Pseudonymous? ☐ Yes ☐ No

If the answer to either of these questions is "Yes," see detailed instructions.

Nature of Authorship Check appropriate box(es). **See instructions**
☐ 3-Dimensional sculpture   ☐ Map   ☐ Technical drawing
☐ 2-Dimensional artwork   ☐ Photograph   ☐ Text
☐ Reproduction of work of art ☐ Jewelry design ☐ Architectural work

3

a **Year in Which Creation of This Work Was Completed**

This information must be given ___ Year in all cases.

b **Date and Nation of First Publication of This Particular Work**
Complete this information ONLY if this work has been published.
Month _____ Day _____ Year _____
_____ Nation

4

See instructions before completing this space.

COPYRIGHT CLAIMANT(S) Name and address must be given even if the claimant is the same as the author given in space 2. ▼

Transfer If the claimant(s) named here in space 4 is (are) different from the author(s) named in space 2, give a brief statement of how the claimant(s) obtained ownership of the copyright. ▼

APPLICATION RECEIVED

ONE DEPOSIT RECEIVED

TWO DEPOSITS RECEIVED

FUNDS RECEIVED

DO NOT WRITE HERE OFFICE USE ONLY

MORE ON BACK ▶ • Complete all applicable spaces (numbers 5-9) on the reverse side of this page.
• See detailed instructions.   • Sign the form at line 8.

DO NOT WRITE HERE
Page 1 of _____ pages

EXAMINED BY

CHECKED BY

☐ CORRESPONDENCE
Yes

FORM VA

FOR
COPYRIGHT
OFFICE
USE
ONLY

DO NOT WRITE ABOVE THIS LINE. IF YOU NEED MORE SPACE, USE A SEPARATE CONTINUATION SHEET.

PREVIOUS REGISTRATION Has registration for this work, or for an earlier version of this work, already been made in the Copyright Office?

☐ **Yes** ☐ **No** If your answer is "Yes," why is another registration being sought? (Check appropriate box.) ▼

a. ☐ This is the first published edition of a work previously registered in unpublished form.

b. ☐ This is the first application submitted by this author as copyright claimant.

c. ☐ This is a changed version of the work, as shown by space 6 on this application.

If your answer is "Yes," give: **Previous Registration Number** ▼   **Year of Registration** ▼

5

DERIVATIVE WORK OR COMPILATION Complete both space 6a and 6b for a derivative work; complete only 6b for a compilation.

a. Preexisting Material Identify any preexisting work or works that this work is based on or incorporates. ▼

b. Material Added to This Work Give a brief, general statement of the material that has been added to this work and in which copyright is claimed. ▼

6

a

b

See instructions
before completing
this space.

DEPOSIT ACCOUNT If the registration fee is to be charged to a Deposit Account established in the Copyright Office, give name and number of Account.

Name ▼   **Account Number** ▼

7

a

CORRESPONDENCE Give name and address to which correspondence about this application should be sent. Name/Address/Apt/City/State/ZIP ▼

b

Area code and daytime telephone number   (   )   Fax number   (   )

Email

CERTIFICATION* I, the undersigned, hereby certify that I am the

check only one ▶ {

☐ author
☐ other copyright claimant
☐ owner of exclusive right(s)
☐ authorized agent of

8

Name of author or other copyright claimant, or owner of exclusive right(s) ▲

of the work identified in this application and that the statements made by me in this application are correct to the best of my knowledge.

Typed or printed name and date ▼ If this application gives a date of publication in space 3, do not sign and submit it before that date.

Date

Handwritten signature (X) ▼

X

**Certificate
will be
mailed in
window
envelope
to this
address:**

Name ▼

Number/Street/Apt ▼

City/State/ZIP ▼

YOU MUST:
• Complete all necessary spaces
• Sign your application in space 8

**SEND ALL 3 ELEMENTS
IN THE SAME PACKAGE:**
1. Application form
2. Nonrefundable filing fee in check or money order payable to *Register of Copyrights*
3. Deposit material

MAIL TO:
Library of Congress
Copyright Office
101 Independence Avenue, S.E.
Washington, D.C. 20559-6000

Fees are subject to
change. For current
fees, check the
Copyright Office
website at
www.copyright.gov,
write the Copyright
Office, or call
(202) 707-3000.

9

*17 U.S.C. § 506(e): Any person who knowingly makes a false representation of a material fact in the application for copyright registration provided for by section 409, or in any written statement filed in connection with the application, shall be fined not more than $2,500.

Instructions for Short Form VA

For pictorial, graphic, and sculptural works

USE THIS FORM IF—

1. You are the **only** author and copyright owner of this work, *and*
2. The work was **not** made for hire, *and*
3. The work is completely new (does not contain a substantial amount of material that has been previously published or registered or is in the public domain).

If any of the above does not apply, you must use standard Form VA.
NOTE: *Short Form VA is not appropriate for an anonymous author who does not wish to reveal his or her identity.*

HOW TO COMPLETE SHORT FORM VA

- Type or print in black ink.
- Be clear and legible. (Your certificate of registration will be copied from your form.)
- Give only the information requested.

NOTE: You may use a continuation sheet (Form __/CON) to list individual titles in a collection. Complete Space A and list the individual titles under Space C on the back page. Space B is not applicable to short forms.

1 Title of This Work

You must give a title. If there is no title, state "UNTITLED." If you are registering an unpublished collection, give the collection title you want to appear in our records (for example: "Jewelry by Josephine, 1995 Volume"). Alternative title: If the work is known by two titles, you also may give the second title. If the work has been published as part of a larger work (including a periodical), give the title of that larger work instead of an alternative title, in addition to the title of the contribution.

2 Name and Address of Author and Owner of the Copyright

Give your name and mailing address. You may include your pseudonym followed by "pseud." Also, give the nation of which you are a citizen or where you have your domicile (i.e., permanent residence).

Please give daytime phone, fax numbers, and email address, if available.

3 Year of Creation

Give the latest year in which you completed the work you are registering at this time. A work is "created" when it is "fixed" in a tangible form. Examples: drawn on paper, molded in clay, stored in a computer.

4 Publication

If the work has been published (i.e., if copies have been distributed to the public), give the complete date of publication (month, day, and year) and the nation where the publication first took place.

5 Type of Authorship in This Work

Check the box or boxes that describe your authorship in the material you are sending. For example, if you are registering illustrations but have not written the story yet, check only the box for "2-dimensional artwork."

6 Signature of Author

Sign the application in black ink and check the appropriate box. The person signing the application should be the author or his/her authorized agent.

7 Person to Contact for Rights/Permissions

This space is optional. You may give the name and address of the person or organization to contact for permission to use the work. You may also provide phone, fax, or email information.

8 Certificate Will Be Mailed

This space must be completed. Your certificate of registration will be mailed in a window envelope to this address. Also, if the Copyright Office needs to contact you, we will write to this address.

9 Deposit Account

Complete this space only if you currently maintain a deposit account in the Copyright Office.

MAIL WITH THE FORM

- A $30 filing fee (Copyright Office fees are subject to change. For current fees, please check the Copyright Office website at *www.copyright.gov*, write the Copyright Office, or call (202) 707-3000.) in the form of a check or money order *(no cash)* payable to "Register of Copyrights," **and**
- One or two copies of the work or identifying material consisting of photographs or drawings showing the work. See table (right) for the requirements for most works. **Note:** Request Circular 40a for more information about the requirements for other works. Copies submitted become the property of the U.S. Government.

Mail everything **(application form, copy or copies, and fee)** *in one package* to:

Library of Congress
Copyright Office
101 Independence Avenue, S.E.
Washington, D.C. 20559-6000

QUESTIONS? Call (202) 707-3000 [TTY: (202) 707-6737] between 8:30 a.m. and 5:00 p.m. eastern time, Monday through Friday. For forms and informational circulars, call (202) 707-9100 24 hours a day, 7 days a week, or download them from the Copyright Office website at *www.copyright.gov*. Selected informational circulars but not forms are available from Fax-on-Demand at (202) 707-2600.

If you are registering:	And the work is *unpublished/published* send:
• 2-dimensional artwork in a book, map, poster, or print	**a.** And the work is *unpublished*, send one complete copy or identifying material **b.** And the work is *published*, send two copies of the best published edition
• 3-dimensional sculpture, • 2-dimensional artwork applied to a T-shirt	**a.** And the work is *unpublished*, send identifying material **b.** And the work is *published*, send identifying material
• a greeting card, pattern, commercial print or label, fabric, wallpaper	**a.** And the work is *unpublished*, send one complete copy or identifying material **b.** And the work is *published*, send one copy of the best published edition

PRIVACY ACT ADVISORY STATEMENT Required by the Privacy Act of 1974 (P.L. 93-579)
The authority for requesting this information is title 17 U.S.C., secs. 409 and 410. Furnishing the requested information is voluntary. But if the information is not furnished, it may be necessary to delay or refuse registration and you may not be entitled to certain relief, remedies, and benefits provided in chapters 4 and 5 of title 17 U.S.C.
The principal uses of the requested information are the establishment and maintenance of a public record and the examination of the application for compliance with the registration requirements of the copyright law.
Other routine uses include public inspection and copying, preparation of public indexes, preparation of public catalogs of copyright registrations, and preparation of search reports upon request.
NOTE: No other advisory statement will be given in connection with this application. Please keep this statement and refer to it if we communicate with you regarding this application.

Copyright Office fees are subject to change. For current fees, check the Copyright Office website at *www.copyright.gov*, write the Copyright Office, or call (202) 707-3000.

Short Form VA
For a Work of the Visual Arts
UNITED STATES COPYRIGHT OFFICE

REGISTRATION NUMBER

VA   VAU

Effective Date of Registration

Application Received

Deposit Received
One   Two

Examined By

Correspondence ☐

Fee Received

TYPE OR PRINT IN BLACK INK. DO NOT WRITE ABOVE THIS LINE.

1 Title of This Work:

Alternative title or title of larger work in which this work was published:

2 Name and Address of Author and Owner of the Copyright:

Nationality or domicile:
Phone, fax, and email:

Phone (   )   Fax (   )
Email

3 Year of Creation:

4 If work has been published, Date and Nation of Publication:

a. Date _____ _____ _____ *(Month, day, and year all required)*
Month   Day   Year

b. Nation

5 Type of Authorship in This Work:
Check all that this author created.

☐ 3-Dimensional sculpture   ☐ Photograph   ☐ Map
☐ 2-Dimensional artwork   ☐ Jewelry design   ☐ Text
☐ Technical drawing

6 Signature:

Registration cannot be completed without a signature.

*I certify that the statements made by me in this application are correct to the best of my knowledge.** Check one:

☐ Author   ☐ Authorized agent

X _____

OPTIONAL

7 Name and Address of Person to Contact for Rights and Permissions:
Phone, fax, and email:

☐ Check here if same as #2 above.

Phone (   )   Fax (   )
Email

8

Certificate will be mailed in window envelope to this address:

Name ▼

Number/Street/Apt ▼

City/State/ZIP ▼

Complete this space only if you currently hold a Deposit Account in the Copyright Office.

9

Deposit Account # _____

Name _____

_____

_____

DO NOT WRITE HERE   Page 1 of _____ pages

*17 U.S.C. § 506(e): Any person who knowingly makes a false representation of a material fact in the application for copyright registration provided for by section 409, or in any written statement filed in connection with the application, shall be fined not more than $2,500.

Trademarks

The most commonly known type of intellectual property is a trademark because we encounter them every day. Popular trademarks such as Microsoft®, Coca-Cola®, Listerine®, Nike®, McDonald's®, and many others are part of life at home, office, or in public. These trademarks have become powerful forms of intellectual property protection with each being worth hundreds of millions, even billions, of dollars.

Trademarks may also be important in protecting certain aspects of industrial designs concepts. Although trademarks will not protect the idea or concept itself, they can protect a proprietary name or symbol or package design used with goods or services. Often the success of a product will depend more on its name and packaging than on the product itself.

The main function of a trademark is to indicate the origin or source of goods. A service mark is like a trademark except that it identifies services instead of goods. A trademark or service mark may also function to symbolize or guarantee the quality of products or services that bear the mark. A trademark is any work, name, symbol, device, or any combination thereof adopted and used by a manufacturer or merchant to identify his or her goods and to distinguish them for those manufactured or sold by others. Rights in a trademark are acquired only by use of the trademark on particular goods (although federal law allows the filing of intent-to-use applications). This right to use a trademark is a property interest that the trademark owner can assert to prevent others from using the mark or one which is confusingly similar.

Federal registration of a trademark is not mandatory, but this registration does provide some major benefits to the trademark owner. Trademarks can also be registered with the appropriate state office for state protection, although even an unregistered trademark can have protection under the common law simply because the mark is used in commerce. To be eligible for federal registration, a trademark must actually be used in commerce, or the owner must have a bona fide intent to use the trademark in commerce. The intent to use the trademark can be extended for periods of six months up to a maximum of two years and is a good way for reserving a trademark with the U.S. Patent and Trademark Office. A trademark is deemed to be used in commerce when it is placed in any manner on goods or their containers or on tags or labels affixed thereto, and the goods are sold or transported in interstate commerce.

Federal registration of a trademark protects the exclusive right of the trademark owner to use the trademark nationwide. Federal registration also gives the trademark owner the right to use the registration symbol ®, which may deter others from using the trademark. Additionally, the trademark owner has the right to sue unauthorized users of the trademark for an injunction, damages, or recovery of profits. If the trademark has not been registered, the designer should use the symbol ™ to protect his or her trademark rights under state or common law.

The U.S. trademark laws provide for the registration of trademarks on two types of registers, designated as the Principal Register and the Supplemental Register. Trademarks that are created, arbitrary, or fanciful can (if otherwise qualified) be easily registered on the Principal Register. However, a trademark that does not qualify for registration on the Principal Register may be registered on the Supplemental Register provided that the trademark applicant shows that the mark is capable of distinguishing applicant's goods and normally has been used in commerce for at least one year.

If all other requirements are satisfied, a trademark may be registered on the Principal Register unless it consists of a mark which: 1) when applied to the applicant's goods or services is merely descriptive or deceptively misdescriptive of them; 2) when applied to the applicant's goods or services is primarily geographically descriptive or deceptively misdescriptive; or 3) is primarily a surname.

As an exception to the general rules, marks may be registered on the Principal Register if they have become distinctive as applied to the applicant's goods in commerce. This usually requires proof of exclusive and continuous use of the mark by the applicant in commerce for the prior five years. A trademark cannot be registered if it consists or comprises of:

1. Immoral, deceptive, or scandalous matter

2. The flag or coat of arms or other insignia of any country or state

3. A name, portrait, or signature identifying a particular living individual (except by his or her written consent)

4. A mark which so resembles a mark registered in the Patent and Trademark Office, or previously used in the U.S. by another and not abandoned, as to be likely when applied to the applicant's goods or services to cause confusion or to cause mistake or to deceive

The designer who believes that a certain product should be trademarked will want to be sure that any chosen trademark does not infringe an already-registered trademark in commercial use. This can be accomplished by conducting a trademark search prior to using and filing an application to register the trademark. If the designer's selected trademark is too similar to other marks in use, a decision should be made to select different trademark.

To conduct a trademark search, the designer can 1) go online to the Web site of the U.S. Patent and Trademark Office (*www.uspto.gov*) and use the Trademark Electronic Search System; 2) go to the public search library at the U.S. Patent and Trademark Office; or 3) use the CD-ROMs containing the trademark database at the patent and trademark depository libraries (a list of these libraries is available from the U.S. Patent and Trademark Office). While the trademark search and application can be filed without the assistance of a search

service or attorney, the evaluation of the trademark search and the official actions from the Trademark Office may require such expert assistance.

The owner of a trademark used or intended to be used in commerce may register the trademark by filing an application for registration in the U.S. Patent and Trademark Office. The application must be filed in the name of the trademark owner and comprises 1) a written application; 2) a drawing of the mark; 3) three specimens or facsimiles of the mark; and 4) the filing fee ($335 for registration in each class at the time of this printing). The application must be in the English language and must specify the name, citizenship, domicile, and post office address of the applicant. A sample trademark application form is provided at the end of this chapter.

After the trademark application has been examined, the U.S. Patent and Trademark Office will allow the registration if all requirements have been met, and the mark is not confusingly similar to other marks. If the application has indicated a bona fide intent to use the mark, the applicant must normally file a statement of actual use (along with dates of actual use, three specimens and the required fee) within six months after receiving a notice of allowance from the Patent and Trademark Office. After allowance, the mark is published for opposition and, if no opposition is filed, the mark is registered.

Trademarks can last forever if the designer keeps using the mark in commerce to identify goods or services. The federal trademark registration now has a term of ten years, but can be renewed for additional ten-year terms. The designer should note that to avoid having the trademark registration canceled, an affidavit of use must be filed with the U.S. Patent and Trademark Office between the fifth and sixth year. Trademarks are entitled to protection in foreign countries under treaties executed by the United States.

Distinctive packaging or trade dress can also provide proprietary rights related to a product. The same is true for unique advertising slogans, colors, jingles, architecture or business designs. The use of these can give the user offensive rights to stop others from using similar techniques which are likely to cause confusion with his or her products or services in the marketplace. Trade dress claims arise under section 43(a) of the federal Lanham Act. A plaintiff must prove three elements to win a trade dress claim: 1) that the features of the trade dress are primarily nonfunctional (i.e., that these features primarily identify the source of the goods or services); 2) that the trade dress has secondary meaning because the public has come to identify the trade dress with the source of the goods or services; and 3) that the competing products' respective trade dresses are confusingly similar, and give rise to a likelihood of confusion among customers as to their sources. When designing a new product or service, the designer should always try to make the product name, packaging, and trade dress as unique and distinctive as possible. This uniqueness gives these features secondary meaning, which is important in acquiring and enforcing trademark and unfair competition rights.

Additional information on trademarks is available in *The Trademark Guide: A Friendly Handbook to Protecting and Profiting from Trademarks*, Second Edition, by Lee Wilson (Allworth Press, 2004). Trademark registration forms and information are also available from the U.S. Patent and Trademark Office, including a pamphlet titled "Basic Facts about Trademarks," which can be downloaded from the U.S. Patent and Trademark Office Web site or requested from the Assistant Commissioner for Trademarks.

Filling in the Form

Identify the mark for which registration is sought. If the mark comprises only words or phrases in non-stylized form, fill these in as capital letters. If the mark has any special form, style or logo, show this in the drawing attached to the application.

Fill in the international class (if known) of the goods or services for which the mark is used. Fill in the name and address of the applicant. Check the appropriate box to indicate whether the applicant is an individual, partnership, corporation, or other entity. Supply the requested information on the applicant's citizenship, domicile, general partners, and state or country of incorporation. Describe the goods or services for which applicant requests registration of the trademark. Attach a copy of the required drawing of the mark to the application.

Check the appropriate box(es) to indicate the basis for the application. Indicate whether the applicant is using the mark in commerce or has a bona fide intention of using the mark in commerce in connection with the goods or services.

If the applicant is using the mark in commerce, state the date of first use of the mark by the applicant anywhere, date of first use of the mark by the applicant in commerce in the U.S., the type of commerce, and the manner or mode of use of the mark in commerce on or in connection with the goods or services. Also include with the application three specimen showing the mark as used in commerce.

If applicant has a bona fide intention to use the mark, then specify the intended manner or mode of use of the mark on or in connection with the goods or services. Also check the appropriate box to show whether the applicant is claiming priority based on a foreign trademark application or registration, and indicate the country, filing date, and registration number. Have the applicant sign and date the trademark application.

International Classification of Goods and Services

While it is not mandatory that the trademark applicant fill in the class of goods and services, the applicant should give proper consideration to which class(es) the goods or services fall within and how to best protect the trademark. This may require registration of the mark in more than one class and selecting classes, such as the services, for broader trademark protection. Some classes of goods and services that would be especially relevant to designs in commerce are:

Goods:

Class 7. Machines and machine tools . . .

Class 8. Hand tools and instruments; cutlery; forks and spoons; razors . . .

Class 9. Scientific, nautical, photographic and electrical apparatus and instruments . . .

Class 10. Surgical, medical, dental and veterinary instruments and apparatus . . .

Class 11. Apparatus for lighting, heating, refrigerating, drying, water supply, sanitary purposes . . .

Class 12. Vehicles, apparatus for locomotion by land, air, or sea . . .

Class 14. Precious metals, jewelry, precious stones, chronometric instruments . . .

Class 15. Musical instruments . . .

Class 16. Paper, cardboard, articles of paper or cardboard, printed matter, photographs . . .

Class 18. Leather articles, skins, hides, traveling bags, umbrellas . . .

Class 20. Furniture, mirrors, picture frames, articles of wood, bone, ivory, plastics . . .

Class 21. Household utensils, containers of non-precious metals, glassware, porcelain, earthenware . . .

Class 24. Textile and textile goods, bed and bed covers . . .

Class 25. Clothing, footwear, headgear . . .

Class 26. Lace, embroidery, ribbons, buttons, pins, needles, artificial flowers . . .

Class 27. Carpets, rugs, mats, linoleums, floor coverings, non-textile wall hangings . . .

Class 28. Games and playthings, gymnastic and sporting articles (except clothing), ornaments, decorations for Christmas trees

Services:

Class 35. Advertising and business

Class 37. Construction and repair

Class 38. Communication

Class 39. Transportation and storage

Class 40. Material treatment

Class 41. Education and entertainment

Checklist for Registering a Trademark

❏ Select a trademark for the goods or services that is fanciful, impressive, and registrable.

❏ Conduct a trademark search for similar marks to 1) avoid infringing other parties' registered trademarks, 2) ascertain that the mark can be registered, and 3) find unregistered marks that would lessen the scope of protection.

❏ Begin using the mark in interstate commerce as early as possible, and file an application for trademark registration promptly.

❏ Make sure that the trademark application includes 1) a completed and written application form, 2) a drawing of the mark, 3) three specimen of the mark if used in commerce, and 4) the filing fee ($335 for registration in each class at the time of this printing).

❏ Complete the trademark application in the English language, and specify the name, citizenship, domicile, and post office address of the applicant. Indicate whether the applicant is an individual, partnership, corporation, or other entity.

❏ Describe the goods or services for which the mark is used or intended to be used.

❏ If known, identify the international class that best describes the goods or services.

❏ Make sure that the drawing is a substantially exact representation of the mark as actually used or to be used in connection with the goods or services. Note that, if the mark is only a word, letter, numeral, or any combination thereof, not depicted in special form, the drawing may be the mark typed in capital letters on paper. Prepare the drawing with care because the rules prohibit any material change to the drawing of the mark after filing.

❏ Follow these specifications for the drawing: The drawing must be made upon pure white durable paper that has a smooth surface. The size of the sheet on which the drawing is made must be 8.5" × 11" long.

The size of the mark must be such as to leave a margin of at least one inch on the sides, top and bottom of the page, and at least one inch between the drawing and the heading. The drawing cannot be more than four inches high and four inches wide. Across the top of the drawing, beginning one inch from the top edge and not exceeding a third of the sheet, there must be a heading which lists on separate lines, applicant's complete name, applicant's address, the goods and services specified in the application, and (if applicable) the date of first use of the mark anywhere and the date of first use of the mark in interstate commerce. Do not include nontrademark matter in the drawing, such as informational matter that may appear on the label.

❑ Indicate that the applicant has adopted and is actually using the mark shown in the drawing which accompanies the application, or that the applicant has a bona fide intention to use the mark, in interstate commerce.

❑ If filing an intent-to-use trademark application, claim priority based upon any foreign trademark application or registration.

❑ Make sure that the application is signed and verified (sworn to) or includes a declaration by the applicant or by a member of the firm or an officer of the corporation or association that is applying for registration of the trademark.

❑ Have the trademark application filed by the owner or by an attorney or other person authorized to practice before the U.S. Patent and Trademark Office.

❑ File the trademark application by mailing to Commissioner for Trademarks, P.O. Box 1451, Alexandria, VA 22313-1451. An application can also be filed online using the Trademark Electronic Application System (instructions available at *www.uspto.gov*).

❑ Respond to any official actions from the Trademark Office within the time specified for a response.

❑ Receive a notice of allowance after the Trademark Office has examined the application and found it acceptable.

❑ If the application is an intent-to-use application, file a statement of actual use (along with dates of actual use, three specimen and the required fee) within six months after receiving the notice of allowance from the Trademark Office (or request additional six-month extensions of time up to a total of three years).

❑ Consult an attorney if the trademark registration involves complex or difficult issues.

Trademark/Service Mark Application, Principal Register, with Declaration

Mark (Identify the mark)

Class No. (If Known)

_____

To the Assistant Commissioner for Trademarks:

Applicant name: _____

Applicant address: _____

_____

Applicant Entity: (Check one and supply requested information)

❑ Individual — Citizenship (Country): _____

❑ Partnership — Partnership domicile (state and country): _____
Names and citizenship (country) of general partners:

_____

_____

_____

_____

❑ Corporation — State (country, if appropriate) of incorporation: _____

_____

❑ Other (Specify nature of entity and domicile): _____

Goods and/or Services:

Applicant requests registration of the above identified trademark/service mark shown in the accompanying drawing in the United States Patent and Trademark Office on the Principal Register established by the Act of July 5, 1946 (15 U.S.C. 1051 et seq., as amended) for the following goods/services:

_____

_____

_____

_____

Basis for Application (Check one or more, but not both the first and second boxes, and supply requested information):

❑ Applicant is using the mark in commerce on or in connection with the above identified goods/services. (15 U.S.C. 105(a), as amended.) Three specimens showing the mark as used in commerce are submitted with this application.

• Date of first use of the mark anywhere: _____

• Date of first use of the mark in commerce which the U.S. Congress may regulate:

_____

• Specify the type of commerce:

_____

(e.g., interstate, between the U.S. and a specified foreign country)

• Specify manner or mode of use of mark on or in connection with the goods/services

_____

(e.g., trademark is applied to labels, service mark is used in advertisements)

❏ Applicant has bona fide intention to use the mark in commerce on or in connection with the above-identified goods/services. (15 U.S.C. 105(b), as amended).
    • Specify intended manner or mode of use of mark on or in connection with the goods or services:

_____

_____

(e.g., trademark will be applied to labels, service mark will be used in advertisements)

❏ Applicant has bona fide intention to use the mark in commerce on or in connection with the above identified goods/services, and asserts a claim of priority based upon a foreign application in accordance with (15 U.S.C. 1126(d), as amended.

    • Country of foreign filing: _____
    • Date of foreign filing: _____

❏ Applicant has a bona fide intention to use the mark in commerce on or in connection with the above identified goods/services and, accompanying this application, submits a certification or certified copy of a foreign registration in accordance with 15 U.S.C. 1126(e), as amended.

    • Country of registration: _____
    • Registration number: _____

Declaration

The undersigned being hereby warned that willful false statements and the like so made are punishable by fine or imprisonment, or both, under 18 U.S.C. 1001, and that such willful false statements may jeopardize the validity of the application or any resulting registration, declares that he/she is properly authorized to execute this application on behalf of the applicant; he/she believes the applicant to be the owner of the trademark/service mark sought to be registered or, if the application is being filed under 15 U.S.C. 105(b), he/she believes applicant to be entitled to use such mark in commerce; to the best of his/her knowledge and belief no other person, firm, corporation, or association has the right to use the above identified mark in commerce, either in the identical form thereof or in such near resemblance thereto as to be likely, when used on or in connection with the goods/services of such other person; to cause confusion, or to cause mistake, or to deceive; and that all statements made of his/her own knowledge are true and all statements made on information and belief are believed to be true.

_____   _____
Date   Signature

_____   _____
Telephone   Print or Type Name and Position

Utility Patents

After the designer has created a proprietary industrial design, it's time to consider protecting the design. If the designer has a patentable idea, he or she should look into trying to protect it under the patent laws. There are two types of patents that might be useful in providing protection for industrial designs— a utility patent and a design patent. This section will focus on utility patents that cover the functional or operational features of an industrial design. Design patents cover the artistic or ornamental features of an industrial design, and they are discussed in the next section.

Whether a utility patent or a design patent will be pursued, the industrial designer should take early steps to document the invention. To document an invention, the designer should get a close friend or associate (who understands the invention, but who is not an inventor thereof) to sign his or her name on a dated diagram or written description of the invention that has also been dated and signed by the designer inventor. The designer can also file a "disclosure document" with the U.S. Patent Office or a provisional patent application. Taking one of these measures will provide evidence of the time that the designer came up with the invention in case of a dispute with other inventors over who conceived it first. Filing a disclosure document with the U.S. Patent Office does not give the designer any patent protection but only provides evidence of the invention.

The designer should make a patent search to see whether or not the invention has already been patented. The designer can make a search on a computerized patent database, such as LEXIS or the U.S. Patent Office Web site (*www.uspto.gov*), or at the U.S. Patent and Trademark Office. The staff at the Patent Office will give the designer some help in conducting patent searches and using Patent Office facilities. If the invention is complex or involves complicated issues, the designer may need the help of a patent agent or patent attorney.

If the invention has not already been patented, the designer can prepare a patent application and file it with the U.S. Patent and Trademark Office. Specific information on preparing and filing a simple U.S. utility patent application is presented later in this section.

A U.S. utility patent can be obtained on a process, machine, article of manufacture, or composition of matter that is 1) new, 2) useful, and 3) unobvious. It can also be a new and useful improvement.

If the invention has been described in a printed publication anywhere in the world, or if it has been in public use or on sale in this country before the date that the designer made the invention, a patent cannot be obtained. That is, it can't already exist. Also, a valid patent cannot be obtained if the invention has been described in a printed publication anywhere, or has been in public use or on sale in this country for more than one year. In this connection, it is immaterial when the invention was made, or whether the printed publication or public use was by the designer or by someone else. The designer must apply for a patent before one year has gone by since the day that he or she described the invention in a printed publication, used it publicly, or placed it on sale; otherwise, any right to a patent will be lost.

The usefulness test is easily met if the designer can show that the invention operates to perform some function. An invention may be useful, although it is destructive. For example, a gun may be patentable subject matter, although its primary purpose is to kill or destroy.

Typically the most difficult hurdle to overcome in establishing patentability is whether the invention is obvious. A patent cannot be obtained even though the invention is not identically disclosed or described in the "prior art" if the differences between the invention and the "prior art" are such that the subject matter "as a whole" would have been "obvious" at the time the invention was made to a person having "ordinary skill in the art" to which the subject matter pertains.

In other words, even if the subject matter sought to be patented is not exactly shown by the prior art, and involves one or more differences over the most similar thing already known, a patent may still be refused if the differences would be obvious. The subject matter sought to be patented must be sufficiently different from what has been used or described before so that it may be said to amount to invention over the prior art. Small advances that would be obvious to a person having ordinary skill in the art are not considered inventions capable of being patented. For example, the substitution of one material for another, or changes in size, are ordinarily not patentable.

If the designer is merely utilizing the teachings or suggestions of the published literature or prior art to solve a problem and no unexpected results are obtained, it is doubtful that the invention overcomes the obviousness test. The invention must be evaluated on the basis of how it relates to the problem faced, the need for a solution, how the invention differs from the prior art, and its prospects of commercial success.

Certain inventions are held not to be patentable because of policy reasons. For example, printed matter cannot be patented, but must be protected by copyright. Inventions useful solely in the utilization of special nuclear material or atomic energy for atomic weapons are excluded from patent protection by the Atomic Energy Act. Remember also that a patent cannot be obtained upon a mere idea or suggestion; rather, a complete description and reduction to practice of the invention are required.

While the designer will probably have to invest considerable time, money, and effort in the invention, the designer can get help from a number of sources. Patent attorneys and agents can help the designer make a patent search and file and prosecute a patent application. Invention promoters are firms that offer—for a fee—to take on the whole job of protecting and promoting the idea. Invention brokers typically work for a portion of the profits from an invention. They may help inventors raise capital and form companies to produce and market the invention. They often provide sophisticated management advice.

Other sources include University Invention/Entrepreneurial Centers, some funded by the National Science Foundation, which provide some help for inventors and innovators. The Small Business Administration's Small Business Institutes (SBIs) are located at several hundred colleges and universities around the country, and they may be able to provide the market research, feasibility analysis, and business planning assistance necessary to make an invention successful. The Office of Energy-Related Inventions in the U.S. Department of Commerce's National Bureau of Standards sometimes evaluates non-nuclear energy-related inventions and ideas for devices, materials, and procedures without charge. Inventor's clubs, associations, and societies are useful sources for networking and gathering information. Talking with other designers and inventors is probably the most helpful thing that a designer can do.

Over the years, Congress, through legislation, and the Courts, through case law in this area have created a set of rules and guidelines that provides a basis for determining what inventions are entitled to patent protection. A designer must evaluate the invention against the required criteria, and if these criteria are met, there is a strong likelihood patent protection of some degree will be available.

Keep in mind that before the designer can have a patentable invention, there must first be conception and reduction to practice. "Conception" means the mental formulation of the invention in sufficient detail that someone familiar with the subject matter to which the invention relates could make and use the invention. Reduction to practice generally involves making or constructing the invention (i.e., preparing a model, diagram, or written description) and testing it to demonstrate its usefulness for its conceived purpose.

Articles of manufacture and machines are the most common types of inventions involving industrial designs. Articles of manufacture include nearly every man-made object from a paper clip to a skyscraper. Machines include any mechanical or electrical apparatus or device (for example, a camera, a bicycle, a computer, an airplane, or the like).

To qualify as new under U.S. law, an invention must:

1. Not have been known, published, or used publicly anywhere by others before the invention was made by the patent applicant

2. Not have been patented or described by anyone in a printed publication anywhere, or on sale in the U.S., more than one year prior to the U.S. filing date of the patent application

3. Not have been abandoned by the patent applicant

4. Not have been first patented in a foreign country, prior to the date of the patent application, based on an application filed more than twelve months before the filing of the U.S. application

5. Not have been described in a patent granted to another where the other patent application was filed in the U.S. before the invention by the patent applicant

6. Not have been made in the U.S. by another before the invention by the patent applicant

The failure of the invention to meet any one of the above criteria means the invention is not novel, but anticipated by the prior art, and bars the right to a U.S. patent.

The disclosure or sale of the invention which will result in a bar to patentability can happen in a number of ways. This may consist of the sale or mere offer of sale of the invention to others, advertising the invention or any other written publication thereof; audio and video disclosure of the invention; and any disclosure of the invention in a speech, journal article, promotional literature, and the like. Any of these acts by the designer or others whereby the invention is disclosed to the public more than one year before filing an appropriate patent application constitutes a bar to a U.S. patent for that invention.

Preparing and Filing a U.S. Utility Patent Application

After reviewing the prior art and determining that the invention is patentable, the designer should file for patent protection on the invention. If the designer is a U.S. inventor, the easiest process is to file a regular or provisional patent application with the U.S. Patent Office. The designer can begin this process by hiring a good patent attorney or agent, or he may be able to prepare and file the patent application himself if he has a good understanding of how to draft a sufficient patent application and get it filed at the U.S. Patent Office.

A complete U.S. utility patent application has several components, some or which are mandatory and others are optional. The mandatory components of a U.S. utility patent application are 1) a specification with one or more claims; 2) drawings, if necessary to describe or "disclose" the invention; 3) the names of the inventors; 4) the required declaration; and 5) the requisite filing fee ($770 at the time of this printing. The fee is reduced to half for small entity—that is, an independent inventor, a small business concern, or a non-profit organization).

The specification, claims, and drawings in a patent application become a part of the granted and published patent and normally follow a specific format. The specification should consist of the following parts:

a) Title

b) Reference to any related patent applications, if any

c) Background of the invention

d) Summary of the invention

e) Brief description of the drawings, if any

f) Detailed description of the invention

g) One or more claims

h) Abstract

A specification with claims, as well as necessary drawings and names of all inventors must be included in the patent application in order for it to be given a filing date. A declaration and filing fee are also mandatory for an application to be complete, but these can be submitted within two months of filing the application without losing the original filing date.

In addition to the above mandatory items, the following optional items should also accompany a completed application: an application transmittal letter, an information disclosure statement listing relevant prior art, a Small Entity Statement (if applicable), and a self-addressed return postcard to acknowledge receipt by the U.S. Patent Office.

The specification is the core of a completed patent application. Before the designer begins to draft the specification it is a good idea to do the following preliminary work to prepare to write the specification:

❏ Become familiar with the patent regulations. There are two primary sets of statutes or regulations that govern all patent matters. These are the Patent Statute, 35 United States Code ("35 U.S.C."); and The Patent Rules, 37 Code of Federal Regulations— a more detailed set of regulations based on the Patent Statute. All of the requirements for drafting an application can be found in the Patent Statute and the Patent Rules. Copies are available from the Government Printing Office (call 202-512-1800 to order them or access the GPO's Web site at *www.access.gpo.gov*).

❏ Write a brief (one- to three-paragraph) description of the invention. Be sure to include all the unique elements of the invention in this description. If there are several variations of the invention, describe all of them.

❏ List all the advantages and benefits of the invention, particularly those that are surprising and unexpected.

❏ If the invention requires drawings in order to fully describe or disclose it, make preliminary sketches, numbering all relevant parts in the drawings. Use as many views as necessary to fully describe and disclose the invention. These sketches are important since they are the basis for the formal drawings that will become part of the application and granted patent.

❏ Review all relevant prior art that the designer has located. The designer can use this art to help come up with terminology and see how similar inventions are described and drawn. Finally, this prior art should be submitted to the Patent Office with an Information Disclosure Statement.

❏ Try drafting a sample claim. A claim is a formalized, precise description of the invention. See the discussion on drafting claims on page 211 for further explanation. This claim should be one sentence that broadly describes the invention and includes all of the necessary elements of the invention.

❏ Review other U.S. patents in the field of the invention to help decide how best to organize the patent application. Since U.S. patents are not protected by copyright law, sections of text from these patents can be copied and included in the designer's patent application.

It is very important that the patent specification meet the requirements of the patent laws, which require basically that:

1. The specification must adequately describe the invention

2. The specification must teach, or "enable," someone else who has skill in the technical area of the invention to make and use the invention

3. The specification must present the best way known by the inventor for practicing the invention

There are a number of reasons why it is imperative that your application meets the requirements of the Patent Statute when you file the application. Firstly, failure to meet these requirements can result in a rejection of your application by the Patent Office or in a subsequent finding of patent invalidity should the granted patent ever be litigated. Secondly, after the patent application has been filed you are not allowed to add new information (called "new matter") without refilling the application. Thirdly, everything you will "claim" as your invention must be in the specification.

Other requirements for the specification are that it be typed on letter-size paper (8.5" × 11"), legal-size paper (8.5" × 13" or 14"), or A4-size paper (21 × 29.7 cm). Spacing should be 1.5 or double-spaced. The specification, claims and abstract should each begin on a new page. Use the following margins for letter and legal-sized paper: at least 1.5" at the top and bottom, and at least 1" on the right and left. Number each page of the specification, preferably at the bottom. Here is a summary of the different parts of the utility patent application specification:

Title Page. The title page should have the title of the invention as well as the names and, if possible, the addresses of all the inventors. It is very important that all inventors be identified. Failure to include the correct inventor(s) could result in a patent being held unenforceable. The title should be descriptive and state in just a few words what is the essence of the invention.

References to Related Applications, if any. If the applicant has previously filed any applications for the same or one or more related inventions, those applications should be referenced at the beginning of the patent specification. For example, the applicant should use language such as:

This application is a [continuation/continuation-in-part/divisional] of application Serial No. ___, filed on _____, entitled "_____."

Omit this part if there are no related applications.

Background of the Invention. This part serves the purposes of providing information about the field of the invention and describing the current state of the art. It typically includes a discussion of the prior art and how the invention is distinguishable over it.

Summary of the Invention. This is a brief synopsis of the invention and its benefits. It can begin by summarizing the objectives identified earlier. Then the subject matter of the invention should be set out in one or more clear, concise sentences or paragraphs. In many cases, the summary of the invention can simply be a recitation of the main claim in the application and a brief explanation of the elements of the invention.

Brief Description of the Drawings. If the application includes drawings, the applicant must provide a brief description of them in the specification. This part of the application typically begins with a sentence such as: "The accompanying drawings further describe the invention." After that introduction, in separate paragraphs, write only one or two sentences describing the drawings, such as "Figure 1 is a perspective view of the machine constructed in accordance with the invention."

Detailed Description of the Invention. This is the most comprehensive part of the specification. In this section of the specification the applicant must fully, clearly, concisely, and exactly describe the invention to enable a technically skilled person to make and use the invention without extensive experimentation. Additionally, the applicant must disclose the "best mode" known by him or her for practicing the invention. Further, the applicant must disclose every element of the invention that the applicant plans to claim.

Abstract. The abstract is presented on a separate sheet of paper at the end of the specification—after the claims. When the patent is published it will appear on the first page. The abstract should be no more than about a 250-word summary of the invention. The purpose of the abstract is to disclose the novel elements of the invention and to help the examiner and the public quickly determine the nature of the technical disclosure. For simplicity, the abstract could be a summary of the main patent claim.

The Claims. The last part of the specification is the claims, which are numbered sentences that define the patentable invention. The claims are perhaps the most important part of the patent application. The claims must describe the invention clearly enough so that anyone reading them knows what is the scope, or the "metes and bounds," of the invention. Following are rules of thumb that should be followed when drafting patent claims:

1. Make sure all claims are supported by the specification. All claimed elements must be described in exactly the same way as they appear in the claims.

2. Use only one sentence per claim. Use commas, semicolons, and colons, but make sure the only period in the claim comes at the very end.

3. Make sure an element has been named once in the claim before modifying or qualifying it. Failure to follow this rule will result in the rejection of the claim for "lack of antecedent basis." More specifically, when naming an element for the first time say: "a rod," or "an insignia." After this first use, the applicant should say "said rod," or "the insignia."

4. Make sure the elements of the claimed invention logically interrelate. A claim must recite an operative combination of elements, not a mere aggregation of elements. The interrelationship between the parts may be structural or it may be functional. Structural relationships between elements are described by words, such as, "connected to," "secured to," "near," "adjacent to," "attached to," or "mixed with." Examples include: "wherein A is connected to B, D is adjacent to C, or E is mixed with F." Functional relationships between elements are described by words such as: "to support," "in order to," "so that it moves." An example is: "wherein A is positioned to support B."

5. Avoid overly precise numbers where possible. If appropriate, use the word "about." The word "about" is acceptable in claims to indicate that a number is not absolutely specific. Examples of the use of "about" in claims include: "comprising about 2 percent water," "heating said liquid for about twenty-five minutes," or "a circumference of no more than about 3 feet." The term "about" helps in obtaining a broader interpretation of the claim.

6. Use functional language if appropriate. Sometimes an element in a claim can best be broadly defined by what it does (e.g., its function). For example, a "means for attaching" could include screws, nails, and adhesives.

7. Be consistent in the use of terms. Once the applicant has used a term in a claim to identify a particular element, he should use that same term consistently throughout the claims. However, define the term as broadly as possible in the description.

8. Avoid unnecessary wordiness. Do not recite in the claims any element unless it is critical to practicing the invention.

9. The applicant should be sure to claim what he or she plans to sell or license. The claims should encompass all possible marketable versions of the invention, including the version that the applicant plans to commercialize.

Drawings. The patent applicant should prepare, or have a draftsman prepare, sketches of the invention. These sketches can be submitted as part of the original filing to meet the drawings requirement. The U.S. Patent Office has many formalities that must be met for formal drawings. Formal drawings must be on a certain size and type of paper, margins must be a certain width, shading and hatching must be used in certain ways, and there are rules for the use of symbols, legends, and arrows. All of these rules can be found in 37 C.F.R. §§ 1.81–1.85 which is provided on the U.S. Patent Office Web site (*www.uspto.gov*).

Declaration. The application must be accompanied by a declaration signed by all the inventors. In this document, the inventor(s) declare(s) he/she/they is/are the sole or joint inventors of the subject matter claimed in the invention. They are also acknowledging a duty to disclose information of which they are aware that is material to the examination of the application. This includes 1) all prior art of which they are aware at the time the application is filed or that they become aware of during the prosecution of the application, and 2) any other information believed to be relevant to a thorough review of the application by the examiner. A sample declaration form is provided at the end of this chapter.

It is important that the declaration not be executed until the application is completed and reviewed by the inventor(s). This rule is strictly maintained by the Patent Office. Therefore, sign the declaration only after the final changes have been made and you have read the final draft of the application.

Filing Fee. There is a basic fee for filing an application with up to three independent claims and a total of twenty claims. If the number of independent and/or dependent claims exceeds this, the fee increases accordingly. Because the fee schedule for patent applications is regularly updated, the applicant should contact the Patent Office before the patent application is to be filed to determine the latest fees. Make the check for the total amount payable to "Commissioner of Patents and Trademarks". Fees are reduced by 50 percent for applicants who qualify as "small entities." At the time of this printing, the regular filing fee for a utility application is $770 and $385 for a small entity. Generally, a small entity is an independent inventor, a business with fewer than 500 employees or a nonprofit organization. If the applicant is filing as a small entity, he or she must also submit a Verified Statement of Small Entity status. A sample Verified Statement of Small Entity Status form can be found on the U.S. Patent Office Web site.

Filing the patent specification, declaration and filing fee meets the statutory requirements for a patent application. However, there are additional documents that the applicant should complete, if appropriate, and file as part of the application with the U.S. Patent Office.

Application Transmittal Letter. The application should include an Application Transmittal Letter addressed to the Commissioner of Patents and Trademarks. This is a cover letter that lists the contents of the package you are sending to the Patent Office, and includes information such as the names of the inventors, the total number of pages of specification and claims, and the number of sheets of drawings. It is also used to calculate the filing fees. See the U.S. Patent Office Web site for a sample Application Transmittal Letter.

Information Disclosure Statement. An applicant for a patent has a legal obligation to disclose to the Patent Office any information that is material to the patentability of the invention. To meet this full disclosure requirement the applicant should file an Information Disclosure Statement (available from the U.S. Patent Office Web site) at the time of filing the application. If the applicant uncovers a piece of prior art after the patent application has been filed, or if he or she discovers that one embodiment of the invention does not work, he or she must advise the Patent Office before the patent issues. Otherwise, if the patent is granted and it is determined that the applicant withheld material information from the Patent Office, a court could rule that the patent is unenforceable.

Certificate of Express Mailing. At the beginning or at the end of the Application Transmittal Letter the applicant should prepare a "Certificate of Express Mailing." This is a signed statement indicating the date that the application was sent by U.S. Express Mail to the U.S. Patent Office. If the patent application is sent to the U.S. Patent Office by Express Mail and it includes a certificate of Express Mailing, then the application will receive the filing date that the application was deposited with the U.S. Postal Service. If the applicant hand delivers the patent application to the U.S. Patent Office, a certificate of express mailing is not necessary. See the sample form at the end of this section.

If the designer has followed all of the above steps, the patent application should be ready for filing with the U.S. Patent Office. Make sure all signatures are in place, no pages or drawings are missing, all documents are in proper form and ready for submission. The specification and all related papers and the filing fee should be placed unfolded in a large envelope addressed to: Box Patent Application, Commissioner for Patents, Washington, D.C. 20231. Alternatively to mailing, the application can be hand carried to the U.S. Patent Office. Remember, if the applicant is going to mail the application, he or she should use only U.S. Express Mail service. In that way the filing date will be the date the application is mailed. Furthermore, if he or she sends the application via U.S. Express Mail there will be proof of the date it was sent (in the form of the Express Mail receipt). Otherwise, the filing date will be the date it is received by the Patent Office. This is true even if you use an overnight delivery service such as Federal Express®.

Instead of a regular utility patent application, the designer can file a less formal provisional application to save initial costs and obtain the earliest filing date. The regular filing fee for a provisional patent application is $160, and this fee is reduced to $80 for a small entity. Like a regular application, a provisional application must adequately and completely describe the invention. But the description does not require all the formalities and sections discussed above for a regular patent application. Before twelve months have passed after filing a provisional application, the applicant can convert it to a regular application.

The review and examination process at the U.S. Patent Office, from filing the application to the grant of a patent, takes about two to three years. When a patent application arrives at the U.S. Patent Office, the applications branch examines the papers to make sure the required parts of the application have been submitted. The application is then assigned a filing date and a serial number. A patent examiner reviews the application for compli-ance with U.S. Patent Office formalities and the Patent Statute and Patent Rules. This review includes a prior art search of earlier patents and publications to determine whether the claims are novel and unobvious. The examiner will then prepare and mail a report called an "office action" setting forth his or her opinion of the application. The office action identifies which claims, if any, are acceptable or "allowed" and which are unacceptable or "rejected," and it gives all the reasons for the rejections. The office action also indicates if there are any other problems with the application. The first office action is usually received about nine to twelve months after the application is filed. The applicant is typically given three months to respond to the office action and this time period can be extended to six months by paying the required fee.

If the examiner finds that the patent claims are patentable and acceptable, the examiner will issue a notice of allowance". After the applicant receives a notice of allowance, all that remains is to take care of any formal matters and to pay the issue fee. At the time of this printing, the regular issue fee for a utility patent is $1,330, which is reduced by half for a small entity.

Filling in the Form

On a separate title page, fill in the title of the invention in item 1, and the complete names of all the inventors in item 2. In item 3, complete any references to related patent applications, including serial numbers and filing dates. In item 4, write a background of the invention, including discussion of the prior art and the problems solved by the invention. In item 5, prepare a brief summary of the invention. Prepare a brief description of the drawings, if any, in item 6. In item 7, write a detailed description of the invention. Starting on a separate page, in item 8, draft the claims covering the invention, with each claim being a separate numbered sentence. Starting on a separate page, write a brief and descriptive abstract of the invention in item 9.

Filling in the Combined Declaration and Power of Attorney Form

In item 2, indicate whether the inventor is sole or joint. Write in a title of the invention. Check the appropriate box(es) to indicate whether the patent application specification is attached, was filed on the specified date and having the identified serial number, or amended on the specified date. In item 4, indicate any prior foreign patent applications on which the inventor is claiming priority benefits. Include application serial numbers, countries, filing dates, and an indication of whether or not priority benefits are being claimed.

In item 5, indicate any prior United States patent applications on which the inventor is claiming priority benefits. Include application serial numbers, filing dates and status. In item 6, indicate the attorney or agent whom the inventor has appointed prosecute the application. Include the telephone number and address to which calls and correspondences should be sent. At the end of the form, fill in the full names of all of the inventors, along with their residences, citizenships, and post office addresses. Have each inventor sign and date the form.

Checklist for Filing a Utility Patent Application

❑ Take steps to maximize the patentability of the invention. Before creating the design, the designer should define the problem that he or she is faced with or the objective you are trying to reach and record this information and his or her ideas. The designer will also need to review prior art to help generate potential solutions and to determine what is new and patentable. After the analysis of the prior art is complete, formulate a research strategy to make the invention. As soon as an idea is conceived of how to solve the problems, write a concise statement of the idea in a notebook. This statement should include all the parameters that are believed to be necessary for solving the problem and as many individual features, elements, compounds, compositions, or process steps that are necessary for the idea to work. Then describe, the plan to proceed to prove whether the idea actually is the answer to solving the problems defined earlier.

❑ Explore all the variables of the invention (i.e., materials of construction, physical parameters, ranges for each component, temperatures, pressures, etc.). In order to obtain the broadest patent coverage, the designer should explore the outer limits of his or her invention. An invention is not completely defined until the inventor knows how to change each of the variables to determine when the invention works and when it does not.

❑ Prepare the patent specification complete with the following components:

   ❑ Title Page, including names of all inventors
   ❑ References to related applications, if any
   ❑ Background of the invention
   ❑ Summary of the invention
   ❑ Brief description of the drawings, if applicable
   ❑ Detailed description of the invention
   ❑ Claims, starting on a separate page
   ❑ Abstract, starting on a separate page

❑ All drawings are present, and each drawing sheet is given a figure number. The inventor's name and address are on the top back of each sheet of formal drawings.

❑ The Declaration is completed, signed and dated in ink by all inventors after having reviewed the patent application specification.

❑ A check or money order payable to the "Commissioner of Patents and Trademarks" for the correct filing fee is enclosed.

❑ If applicable, the Verified Statement of Small Entity Status form is completed, signed, and dated in ink.

❑ An Information Disclosure Statement (Form PTO-1449) is completed. Copies of all cited references are attached.

❏ A properly completed and signed Transmittal Letter is enclosed which identifies every document that is being sent to the U.S. Patent Office. A Certificate of Express Mail is included with the Transmittal Letter.

❏ A self-addressed return receipt postcard, with the mailing date and all papers listed on the back.

❏ Place the complete patent application in a large envelope and send via Express Mail to:

Box Patent Application
Commissioner for Patents
Washington, DC 20231

❏ Respond to any Office Actions from the U.S. Patent Office within the time prescribed or after requesting an extension of time.

❏ Pay the issue fee if a Notice of Allowance and Issue Fee Due is received from the U.S. Patent Office.

Utility Patent Application

(Put title of the invention and names of the inventors together on a separate page)

1. _____ (Title of the Invention)

2. _____ (Names of the Inventors)

3. Reference to related patent applications:

_____
_____
_____
_____

4. Background of the invention:

_____
_____
_____
_____

5. Summary of the invention:

_____
_____
_____
_____
_____

6. Brief Description of the drawings:

_____
_____
_____
_____
_____

7. Detailed description of the invention:

_____
_____
_____
_____
_____
_____

(Start on a separate page)

8.   Claims:

Claim 1. _____
_____
_____
_____

Claim 2. _____
_____
_____
_____

Claim 3. _____
_____
_____
_____

(Start on a separate page)

9.   Abstract:

CERTIFICATE OF EXPRESS MAILING PURSUANT TO 37 C.F.R.§1.10

I hereby certify that this New Application Transmittal and the documents referred to as enclosed herein are being deposited with the United States Postal Service on this date _____ in an envelope bearing "Express Mail Post Office to Addressee" Mailing Number _____ addressed to the Commissioner for Patents, Washington, D.C. 20231.

_____

(Typed or printed name of person mailing)

_____

(Signature of person mailing papers and the date)

Combined Declaration and power of Attorney in Original Application

1. As a below named inventor, I hereby declare that:

i) my residence, post office address, and citizenship are stated below next to my name; and that

ii) I verily believe I am an original, first, and ❑ sole ❑ joint inventor of the improvement in **[Title of Invention]** described and claimed in the specification which
❑ is attached hereto.
❑ was filed on _____ as Application Serial No. _____ and was amended on _____.

2. I hereby state that I have reviewed and understand the contents of the above-identified specification, including the claims, as amended by any amendment referred to above.

3. I acknowledge the duty to disclose information which is material to the examination of this application in accordance with the provisions of 37 C.F.R. § 1.56.

4. I hereby claim foreign priority benefits pursuant to 35 U.S.C. § 119 from any foreign application(s) listed below and have also identified below any foreign application(s) for patent or inventor's certificate which have a filing date prior to that of the application from which priority is claimed.

Prior Foreign Application(s) (if any):

Number	Country	Filing Date	Priority Claimed
_____	_____	_____	_____
_____	_____	_____	_____
_____	_____	_____	_____

5.   I hereby claim the benefit under 35 U.S.C. § 120 from any United States application(s) listed below and, insofar as the subject matter of each of the claims of this application is not disclosed in the prior United States application in the manner provided by the first paragraph of 35 U.S.C. § 112, I acknowledge the duty to disclose material information as defined in 37 C.F.R. § 1.56 which occurred between the filing date of the prior application and the national or PCT international filing date of this application:

Application Serial Number	Filing Date	Status (Patented, Pending, Abandoned)
_____	_____	_____
_____	_____	_____
_____	_____	_____

6.   I hereby appoint the following attorney(s) and/or agent(s) to prosecute this application and to transact all business in the Patent and Trademark Office connected therewith:

Address all telephone calls to _____ at telephone number: _____
Address all correspondence to: _____

7.   I hereby declare that all statements made herein of my own knowledge are true and that all statements made on information and belief are believed to be true, and further, that these statements were made with knowledge that willful false statements and the like are punishable by fine or imprisonment, or both, pursuant to 18 U.S.C. § 1001 and that such willful false statements may jeopardize the validity of the application or any patent issued thereon.

_____
Full name of inventor

_____
Signature of Inventor

Date: _____

Residence: _____

Citizenship: _____

Post Office Address: _____

Design Patents

In addition to utility patents, design patents can be very useful in protecting industrial designs. In general terms, a "utility patent" protects the way an article is used and works, while a "design patent" protects the way an article looks. A design patent protects the new shape or ornamental or artistic features of an article. Unlike a utility patent, a design patent does not protect the functional elements of an invention. If a design has any functionality, then it cannot be protected with a design patent at all, even if the design is also ornamental. Some examples of designs which can be patented are unique bottle shapes, a computer screen display, an artistic lamp, an ornamental pin, a watch design, the grill of an automobile, and the like. Ornamental designs are an important feature of practically every product on the market.

The designer can file for a design patent and a utility patent on the same article. The design patent would be directed to the exterior appearance of the article whereas the utility patent would be directed to its function. The designer may also be able to copyright his or her unique design in addition to obtaining a patent on it. Because one form of patent or intellectual property coverage does not preclude another, the industrial designer should consider all the available options for protecting your invention.

A design patent has a fourteen-year term from the date of grant, and it protects against any design that is substantially similar to the patented design. A design patent is relatively easy to obtain if the design is new and nonfunctional. A design patent application is also simple to prepare and should include a pre- amble, a specification that explains the design, a single claim, one or more drawings, a formal inventor's declaration and the filing fee. A sam- ple design patent application form is presented at the end of this section.

A design patent covers the visual ornamental features that are embodied in, or applied to, an article of manufacture. Because a design patent is based upon appearance, the subject matter of a design patent application may relate to the configuration or shape of an article, to the surface ornamentation applied to an article, or to a combination of these.

The U.S. patent laws provides for the grant- ing of a design patent to any person who has invented any new, original, and ornamental design for an article of manufacture. An ornamental design may be embodied in a por- tion of or in the entire article. Keep in mind that a design patent protects only the appearance of the article and not structural or utilitarian features.

A design for a product, which is required by the function of the product, cannot be covered by a design patent because the design is not ornamental. This means that to obtain a design patent the appearance of the product must not be dictated by the function that the product performs. It is also required that for a design to be patentable, the design must be original. Thus, a design that simulates a known or naturally occurring object or person is not original.

A design patent application must include the following:

1. Preamble, stating name of the applicant, title of the design, and a brief description of the nature and intended use of the article in which the design is embodied

2. Description of the figure(s) of the drawing

3. A single claim

4. Drawings or photographs

5. Executed oath or declaration

6. The required filing fee ($340 at the time of this printing). If the applicant is a small entity, the filing fee and other fees are reduced by half.

A design must be shown in the drawing as applied to an article of manufacture if the design is directed solely to surface ornamentation. The article of manufacture itself forms no part of the design, and the article must be shown in broken lines in the drawings. The preamble should state the name of the design patent applicant, the title of the design, and a brief description of the nature and intended use of the article in which the design is embodied. This information will be printed on the design patent if the patent is issued.

The title of the design patent application should be specific and descriptive. The title must identify the article in which the design is embodied by the name publicly known and used. The title should not contain trademarks or other marketing designations.

Only one patent claim covering the respective design may be included in a design patent application. This is because each ornamental design is a distinct invention and must be filed in a separate design patent application. The claim is important because it defines the design that the applicant wishes to patent, in terms of the article in which it is embodied or applied. The claim has to be presented in the formal language: "The ornamental design for (name of the article which embodies the design or to which it is applied) as shown and described." The design patent applicant should use consistent terminology in the title, preamble, drawings, and patent claim.

The descriptions of the figures indicate what each view of the drawings represents (e.g., front view, top view, side view, rear view, perspective view, etc.). Because the drawings themselves are the best depiction of the design, the brief descriptions of the drawings are the only written explanation of the design that is required in the specification of a design patent application. However, statements or feature descriptions may be included in the design patent application for clarification, such as the following:

❏ A description which disclaims parts of the article that form no part of the claimed design

❏ A description of parts of the claimed design which are not shown in the drawings, such as views which are the same or mirror images of each other

❏ A description which states the nature and use of the claimed design, particularly if this is not included in the preamble

❏ A statement indicating that any broken-line illustration of the background structure in the drawing is not part of the claimed design

The drawings are one of the most important parts of a design patent application. A drawing or a black-and-white photograph of the claimed design must be included in every design patent application. These drawings or photographs of the claimed design should be a clear and complete depiction of the design that is to be patented. The design drawing or photograph must include sufficient views of the design to constitute a complete disclosure of the appearance of the design claimed.

The U.S. Patent Office has specific and detailed requirements for the proper drawings and/or photographs that are included in design patent applications. These requirements are published in chapter 37 of the Code of Federal Regulations Sections 1.84(b)(1) and 1.152. These are available from the Superintendent of Documents (United States Government Printing Office, Washington, D.C. 20402; telephone: 202–512–1800) and at the U.S. Patent Office Web site (*www.uspto.gov*).

The design patent applicant is required to execute an oath or declaration in compliance with the U.S. Patent Office requirements. A sample The "Combined Declaration and Power of Attorney," which is useful for design patent applications, is provided and discussed in the previous section on utility patents.

Helpful information on design patents can be found in "A Guide to Filing a Design Patent Application" at the U.S. Patent Office Web site. Useful information from the U.S. Patent Office Web site on the design patent application process (also applicable to utility patents) is presented here:

Filing an Application

When the Office receives a complete design patent application, along with the appropriate filing fee, it is assigned an Application Number and a Filing Date. A "Filing Receipt" containing this information is sent to the applicant. The applications are then assigned to an examiner. Applications are reexamined in order of their filing date.

Examination

The actual "examination" entails checking for compliance with formalities, ensuring completeness of the drawing disclosure, and a comparison of the claimed subject matter with the "prior art." "Prior art" consists of issued patents and published materials. If the claimed subject matter is found to be patentable, the application will be "allowed," and instructions will be provided to applicant for completing the process to permit issuance as a patent.

The examiner may reject the claim in the application if the disclosure cannot be understood or is incomplete, or if a reference or combination of references found in the prior art shows the claimed design to be unpatentable. The examiner will then issue an Office Action detailing the rejection and addressing the substantive matters which effect patentability.

Response

If, after receiving an Office Action, the applicant elects to continue prosecution of the application, a timely reply to the action must be submitted. This reply should include a request for reconsideration or further examination of the claim, along with any amendments desired by the applicant, and must be in writing. The reply must distinctly and specifically point out the supposed

errors in the Office Action and must address every objection and/or rejection in the action. If the examiner has rejected the claim over prior art, a general statement by the applicant that the claim is patentable, without specifically pointing out how the design is patentable over the prior art, does not comply with the rules.

In all cases where the examiner has said that a reply to a requirement is necessary, or where the examiner has indicated patentable subject matter, the reply must comply with the requirements set forth by the examiner, or specifically argue each requirement as to why compliance should not be required. In any communication with the Office, the applicant should include the following items:

1. Application number (checked for accuracy).
2. Group art unit number (copied from filing receipt or the most recent Office Action).
3. Filing date.
4. Name of the examiner who prepared the most recent Office Action.
5. Title of invention.

It is the applicant's responsibility to make sure that the reply is received by the Office prior to the expiration of the designated time period set for reply. This time period is set to run from the "Date Mailed," which is indicated on the first page of the Office Action. If the reply is not received within the designated time period, the application will be considered abandoned. In the event that applicant is unable to reply within the time period set in the Office action, abandonment may be prevented if a reply is filed within six months from the mail date of the Office action provided a petition for extension of time and the fee set forth in 37 CFR § 1.17(a) are filed. The fee is determined by the amount of time requested, and increases as the length of time increases. These fees are set by Rule and could change at any time.

An "Extension of Time" does not have to be obtained prior to the submission of a reply to an Office Action; it may be mailed along with the reply.

Note: an extension of time cannot be obtained when responding to a "Notice of Allowance."

To ensure that a time period set for reply to an Office Action is not missed, a "Certificate of Mailing" should be attached to the reply. This "Certificate" establishes that the reply is being mailed on a given date. It also establishes that the reply is timely, if it was mailed before the period for reply had expired, and if it is mailed with the United States Postal Service. A "Certificate of Mailing" is not the same as "Certified Mail." A suggested format for a Certificate of Mailing is as follows:

> "I hereby certify that this correspondence is being deposited with the United States Postal Service as first class mail in an envelope addressed to: Box Design, Commissioner for Patents, Washington, D.C. 20231, on (DATE MAILED)"

_____

(Name: Typed or Printed)

_____

Signature

_____

Date

If a receipt for any paper filed in the USPTO is desired, the applicant should include a stamped, self-addressed postcard, which lists, on the message side applicant's name and address, the application number, and filing date, the types of papers submitted with the reply (i.e., one sheet of drawings, two pages of amendments, one page of an oath/declaration, etc.) This postcard will be stamped with the date of receipt by the mailroom and returned to the applicant. This postcard will be applicant's evidence that the Office received the reply on that date. If the applicant changes his or her mailing address after filing an application, the Office must be notified in writing of the new address . . . . The applicant's failure to receive and properly reply to these Office communications will result in the application being held abandoned. Notification of "Change of Address" should be made by separate letter, and a separate notification should be filed for each application.

Reconsideration

Upon submission of a reply to an Office Action, the application will be reconsidered and further examined in view of the applicant's remarks and any amendments included with the reply. The examiner will then either withdraw the rejection and allow the application, or, if not persuaded by the remarks and/or amendments submitted, repeat the rejection and make it final. The applicant may file an appeal with the Board of Patent Appeals and Interferences after being given a final rejection or after the claim has been rejected twice. Applicant may also file a new application prior to the abandonment of the original application, claiming benefit of the earlier filing date. This will allow continued prosecution of the claim."

The preparation, filing, and prosecution of a design patent application with the U.S. Patent Office requires a basic knowledge and understanding of the patent laws, rules, and procedures. Although a knowledgeable applicant can handle the design patent application, the assistance of a patent attorney or patent agent may be needed to obtain the greatest patent protection to which the applicant is entitled.

Filling in the Form

In item 1, include the complete names of all of the inventors. Also include a title of the invention. Include in item 2 a cross-reference to any related patent applications, along with application serial numbers and filing dates. In item 3, include a brief description indicating the type of view (e.g., front, side, rear, top, etc.) shown for each of the figures of drawings. For figure 1, which shows a perspective view, fill in the name of the article that incorporates the design. In item 4, include any additional feature description, such as the nature and use of the claimed design, particularly if this is not included in the preamble. In the patent claim under item 5, fill in the name of the article which incorporates the design.

Checklist for Filing a Design Patent Application

❏ Prepare the complete design patent application arranged with the following components:

  ❏ Design application transmittal form (available from the U.S. Patent Office Web site)

  ❏ Specification. The specification should include the following sections in order:

    ❏ Preamble, stating the name of the applicant, title of the design, and a brief description of the nature and intended use of the article in which the design is embodied

    ❏ Cross-reference to related applications

    ❏ Description of the figures of the drawing

    ❏ Feature description

    ❏ A single claim

  ❏ Drawings or photographs (each drawing sheet is given a figure number, and the inventor's name and address are on the top back of each sheet of formal drawings)

  ❏ Executed oath or declaration

❏ A check or money order payable to the "Commissioner of Patents and Trademarks" for the correct filing fee is enclosed.

❏ If applicable, the Verified Statement of Small Entity Status form is completed, signed, and dated in ink.

❏ An Information Disclosure Statement (Form PTO-1449) is completed. Copies of all cited references are attached.

❏ A Certificate of Express Mail is included with the Transmittal Form.

❏ A self-addressed, return receipt postcard, with the mailing date and all papers listed on the back.

❏ Place the complete patent application in a large envelope and send via Express Mail to:

  Box Design Patent Application

  Commissioner for Patents

  Washington, DC 20231

❏ Respond to any office actions from the U.S. Patent Office within the time prescribed or after requesting an extension of time.

❏ Pay the issue fee if a notice of allowance and issue fee due is received from the U.S. Patent Office.

Design Patent Application

1. Preamble

I, _____[Name(s) of the Inventor(s)]
have invented a new design for a _____
(Title of the Invention) as set forth in the following specification:

_____
_____
_____
_____
_____
_____
_____
_____
_____
_____
_____

2. Cross-references to Related Applications

_____
_____
_____
_____
_____

3. Brief Description of the Drawings

Figure 1 is a perspective view of a _____ [Name of Article] showing my new design;

Figure 2 is a _____ view thereof;
Figure 3 is a _____ view thereof;
Figure 4 is a _____ view thereof; and
Figure 5 is a _____ view thereof.

4. Feature Description

_____
_____
_____
_____

5. Claim

I claim: The ornamental design for _____ [Name of Article] as shown and described.

Selected Bibliography

Battle, Carl W. *The Patent Guide*. New York: Allworth Press, 1997

Borja de Mozota, Brigitte. *Design Management*: *Using Design to Build Brand Value and Corporate Innovation*. New York: Allworth Press, 2003.

Bruce, Margaret, Bessant, J.R. *Design in Business*. New York: Financial Times/Prentice Hall, 2002.

Dreyfuss, Henry. *Designing for People*. New York: Allworth Press, 2003.

Epstein, Lee. *Legal Forms for the Designer*. New York: N. & E. Hellman, 1977.

Gorman, Carma, (ed.). *The Industrial Design Reader*. New York: Allworth Press, 2003.

Leland, Caryn. *Licensing Art and Design*. New York: Allworth Press, 1995.

Sack, Steven Mitchell. *From Hiring to Firing*. Merrick, New York: Legal Strategies Publications, 1995.

Sack, Steven Mitchell. *The Complete Collection of Legal Forms for Employers*. Merrick, New York: Legal Strategies Publications, 1996.

Wilson, Lee. *The Copyright Guide*: *A Friendly Handbook for Protecting and Profiting from Copyrights*, Third Edition. New York: Allworth Press, 2003.

Wilson, Lee. *The Trademark Guide*: *A Friendly Handbook to Protecting and Profiting from Trademarks*, Second Edition. New York: Allworth Press, 2004.

Index

Books from Allworth Press

Allworth Press is an imprint of Allworth Communications, Inc. Selected titles are listed below.

Designing for People
by Henry Dreyfuss (paperback, 6 ¾ × 9 ½, 256 pages, $21.95)

The Industrial Design Reader
edited by Carma Gorman (paperback, 6 × 9, 256 pages, $19.95)

Design Management
by Brigitte Borja de Mozota (paperback, 6 × 9, 256 pages, $19.95)

Creating the Perfect Design Brief
by Peter L. Philips (paperback, 6 × 9, 224 pages, $19.95)

Turn Your Idea or Invention into Millions
by Don Kracke (paperback, 6 × 9, 224 pages, $18.95)

Licensing Art and Design, Revised Edition
by Caryn R. Leland (paperback, 6 × 9, 128 pages, $16.95)

The Interior Designer's Guide to Pricing, Estimating, and Budgeting
by Theo Stephan Williams (paperback, 6 × 9, 240 pages, $19.95)

Citizen Designer: Perspectives on Design Responsibility
edited by Steven Heller and Véronique Vienne (paperback, 6 × 9, 272 pages, $19.95)

Looking Closer 4: Critical Writings on Graphic Design
edited by Michael Bierut, William Drenttel, and Steven Heller (paperback, 6 ¾ × 9 ⅞, 304 pages, $21.95)

Graphic Design Time Line: A Century of Design Milestones
edited by Steven Heller and Elinor Pettit (paperback, 6 ¾ × 9 ⅞, 272 pages, $19.95)

Graphic Design History
edited by Steven Heller and Georgette Balance (6 ¾ × 9 ⅞, 352 pages, $21.95)

Design Issues: How Graphic Design Informs Society
edited by DK Holland (paperback, 6 ¾ × 9 ⅞, 288 pages, $21.95)

Please write to request our free catalog. To order by credit card, call 1-800-491-2808 or send a check or money order to Allworth Press, 10 East 23rd Street, Suite 510, New York, NY 10010. Include $5 for shipping and handling for the first book ordered and $1 for each additional book. Ten dollars plus $1 for each additional book if ordering from Canada. New York State residents must add sales tax.

To see our complete catalog on the World Wide Web, or to order online, you can find us at **www.allworth.com.**